Industrial Cultures and Production
Understanding Competitiveness

AF714888

Springer-Verlag London Ltd.

Lauge Rasmussen and Felix Rauner (Eds)

Industrial Cultures and Production

Understanding Competitiveness

Springer

Lauge Rasmussen
Technical University of Denmark, Institute of Technology & Social Sciences
DK-2800, Lyngby, Denmark

Felix Rauner
Institute of Technology and Education, University of Bremen, Germany

ISBN 978-3-540-76029-0

British Library Cataloguing in Publication Data
Rasmussen, Lauge Baungaard, 1947-
Industrial cultures and production : understanding competitiveness
1.Competition - Cross-cultural studies 2.Technological innovations - Cross-cultural studies
I.Title II.Rauner, Felix
338.6'048
ISBN 978-3-540-76029-0 ISBN 978-1-4471-1492-5 (eBook)
DOI 10.1007/978-1-4471-1492-5

Library of Congress Cataloging-in-Publication Data
A catalog record for this book is available from the Library of Congress

Apart from any fair dealing for the purposes of research or private study, or criticism or review, as permitted under the Copyright, Designs and Patents Act 1988, this publication may only be reproduced, stored or transmitted, in any form or by any means, with the prior permission in writing of the publishers, or in the case of reprographic reproduction in accordance with the terms of licences issued by the Copyright Licensing Agency. Enquiries concerning reproduction outside those terms should be sent to the publishers.

© Springer-Verlag London 1996
Originally published by Springer-Verlag London Limited in 1996

The use of registered names, trademarks etc. in this publication does not imply, even in the absence of a specific statement, that such names are exempt from the relevant laws and regulations and therefore free for general use.

The publisher makes no representation, express or implied, with regard to the accuracy of the information contained in this book and cannot accept any legal responsibility or liability for any errors or omissions that may be made.

Typesetting: Camera ready by editors

34/3830-543210 Printed on acid-free paper

Preface

This book contains a selection of articles on the subject of 'Culture and Production'. They are results of international conferences held in Tokyo, Washington and Bremen between 1991 and 1994. The International Research Network on Culture and Production (CAPIRN) carried out a 5-year joint research project examining the impact of different industrial cultures on the development and implementation, and above all on the international transfer of technology. The machine tools sector was selected for this international comparative study, because over the last 15 years this global market has undergone dramatic changes that cannot be adequately explained by traditional economic theories of international competition. The 'industrial culture' research concept permits an analysis and understanding of hitherto unrecognised interrelationships between the dimensions of different industrial cultures and the process of technological innovation in international competition.

The special challenge faced by CAPIRN was to develop the theoretical concept of industrial culture further and to apply it within a large-scale international study. A considerable amount of work in this field has been published by CAPIRN members since 1990. This book is the first compilation of research findings in the field of industrial culture.

We wish to express our thanks to the national research councils in the participant countries, the FORCE and FAST programmes of the European Union, the Japanese Ministry for Industry, MITI, and the Hans Böckler Foundation, to mention only some of the many bodies that have provided support. The results of industrial culture research signify a further step along the path to greater interdisciplinarity in industrial and technological research. One can only hope that this interdisciplinarity will form a much more significant component in the education and training of engineers, economists, industrial sociologists and occupational psychologists in the future. The complex nature of the problems associated with industrial change demand interdisciplinary and multidisciplinary know-how in companies, governments and industrial agencies. This book is therefore addressed to all those who wish to understand and shape the organisation of technological innovation, and the competition bases on that process.

Lauge Rasmussen
Copenhagen

Felix Rauner
Bremen

Contents

Contributors

Thomas Binder
Engineer, Consultant. Department of Human Resources Development
Danish Technological Institute, Denmark

Peter Brödner
Engineer, Director of Wissenschaftszentrum, NRW
Institute of Work and Technology, Gelsenkirchen, Germany

J. Martin Corbett
Psychologist, Senior Lecturer. Warwick Business School
University of Warwick, UK

Jürgen Friedrich
Engineer, Professor of Computer Science Department
University of Bremen, Germany

Yoshimi Ito
Engineer, Professor of Department of Mechanical Engineering for Production
Tokyo Institute of Technology, Japan

Gabrielle Laske
Psychologist, Researcher. Institute of Technology and Education
University of Bremen, Germany

Eckehard Fozzy Moritz
Engineer, Researcher. Institute of Technology and Education
University of Bremen, Germany

Klaus T. Nielsen
Sociologist, Researcher. Department of Environment,
Technology and Social Studies, Roskilde University, Denmark

Lauge Rasmussen
Sociologist, Senior Lecturer. Institute of Social Sciences
Technical University of Denmark

Felix Rauner
Engineer, Professor of Institute of Technology and Education
University of Bremen, Germany

Klaus Ruth
Sociologist, Researcher. Institute of Technology and Education
University of Bremen, Germany

Introduction: Approaches to Understanding Culture

by Lauge Rasmussen and Felix Rauner

Three research directions and lines of discussion concerned with questions of organization of innovation take up the term culture and attempt to make it fruitful for empirical research: industrial culture, economic culture and corporate culture.

Ever since the rise of Japan as a leading industrial power it has been popular to discuss organizations in cultural terms. The original impetus behind "corporate culture" was to counter the lack of motivation and identification in many big, Western Tayloristic oriented companies. Thus "corporate culture" is a management concept rather than a scientific approach. Nevertheless, organization literature has been very concerned to find a scientific interpretation of social-cultural qualities that develop within organizations.

The concept of culture in much of this literature is limited. Culture is conceived of as an internal attribute of the company rather than an external factor that impinges on the company. Organizations are perceived as mini-societies that have their own distinctive patterns of culture and subculture. Drawing upon anthropology's structural functionalistic conception of culture, much corporate culture literature implies that ceremonies and rituals contributes in some way to the effectiveness of an organization. Thus, a strong internal corporate culture is believed to be a possession that gives a company a competitive advantage (Barney, 1986; Kilmann et al, 1985). In their "search for excellence" in American corporations Thomas Peters and Robert Waterman place considerable emphasis on the importance of positive reinforcement (the reward for desirable behaviour). (Peters, Waterman, 1982). Viewed as one metaphor among others of the social life in small and big organizations the cultural metaphor may open the scope of understanding the complex relationships between formal and informal patterns of shared meanings and symbolic behaviour. As Gareth Morgan points out:

"The cultural metaphor points towards another means of creating organized activity: by influencing the language, norms, folklore, ceremonies, and other social practices that communicate the key ideologies, values, and beliefs guiding action" (Morgan, 1986, p. 135).

Nevertheless, corporate culture literature tends to ignore the cultural-historical circumstances that allow for example, Japanese companies to flourish as they do. Because the concept of corporate culture is limited to the internal life of the company, it tends to overestimate the ease with which techniques and organizational behaviour can be transplanted from one context to another. It is exactly the context that often makes the difference between "success" and "failure".

This point has been somewhat better handled by another approach of culture.

The contours of the concept of economic culture worked up by Hofstede (1980) and Berger (1987) have been formulated relatively clearly. Different cultural configurations of the state or of large economic regions are assumed. There is a constitutive assumption that different economic cultures are reflected in the performance of national or regional economies. Thus Bond and Hofstede, for example, examine the connection between culture and economy by distinguishing between four main dimensions of culture: "...power distance, individualism, masculinity versus femininity and uncertainty avoidance..." (p. 5) and translate them into a standardized empirical set of instruments. They compare 53 countries according to their GNP and economic growth. They were, however, not able to identify a definite connection between economic culture and the comparative macroeconomic factors. The concept of Confucian cultural ethics, based on Weber's theory of the Protestant ethic, turned out to be more successful. Bond and Hofstede distinguished between "... a Western transcendentalist ethic concerned with the search for Truth (uncertainty avoidance) whose focus on analysis, rather than... the Confucian value, is in the present world an economic handicap" (p. 5).

The widespread rediscovery of what Peter L. Berger (1987) has termed "economic culture" is "...not a theory within the science of economics; rather, it is an intellectual exercise on the borderline of economics and other social sciences. An "economic culture" theory of capitalism (or for that matter, of any other economic phenomenon) will explore the social, political and cultural matrix on context within which these particular economic processes operate" (Berger, 1987, p. 10,7). In difference to the concept of "corporate culture" Berger's focus is more on context variables than on internal attributes of the company. He is very concerned "...to avoid the distinctions of so-called convergence theory, which trivializes very important economic, social and political differences." (Berger, 1987, p. 31). According to Berger we are in a "...global laboratory in which the "chemical reaction" of modernization may be observed in a series of more or less complete experiments.

The three crucial test tubes represent Western industrial capitalism, East Asian industrial capitalism, and industrial socialism; the additional test tubes represent various Third World societies." (Berger, 1987, p. 4).

The problem with the concept of economic culture is that although the connection between culture and economic success can be plausibly explained with the help of rough categories, the exact connection, the question of the steps for transmission of these connections, remains in the dark. The term culture is also used in the concept of "economic culture" for derived terms like "management culture". A number of authors warn against a too sweeping reference to concepts like "post-Confucianism" as the point that explains everything about the success of Far Eastern economic cultures. The obvious necessity to distinguish between the economies of Japan, China and the NICs results in a differentiation of the concept of economic culture.

While Berger's general concept of "economic culture" designates a macro level of culturally bounded characteristics, industrial culture refers to more specifically meso-cultural phenomena. One of these are the attitudes to state interventions held by key institutional elites. For example Dyson and Wilks (1983) use the term "industrial culture" arguing that "...countries have evolved their own distinctive industrial cultures which reflect traditions of public authority but are also moulded by the historical conditions of industrialization. Such industrial cultures are expressed in attitudes and behaviour but are also embodied in institutions and form a mediating factor in the policy process" (Wilks, 1990, p. 131).

Such attitudes are taken to be tacit but specific guides to action embodied in patterns of organizational norms and influencing processes of financing, planning and executing activities as well as recruitment of new members.

Wilks contends that the origins, form and effect of culture is "...part of a process of innovation." (Wilks, 1990, p. 147). A similar approach of understanding forces behind innovation in different industrial contexts is taken by Michael E. Porter in his book "The Competitive Advantage of Nations": "...Competitive advantage is created and sustained through a highly localized process. Differences in national economic structures, values, cultures, institutions, and histories contribute profoundly to competitive success. The role of the home nation seems to be as strong as or stronger than ever. While globalization of competition might appeal to make the nation less important, instead it seems to make it more so. With fewer impediments to trade to shelter incompetitive domestic firms and industries, the home nation takes on growing significance because it is the source of the skills and technology that underpin competitive advantage." (Porter, 1990, p. 19). Following the tradition of Joseph Schumpeter, Porter tries to explain "...how a nation provides an environment in which its firms are able to improve and innovate faster than foreign rivals in a particular industry." (Porter, 1990, p. 20). Porter's well known "diamond" model of determinants of national advantage ("factor conditions", "demand conditions", "related and supporting industries" and "firm strategy, structure and rivalry") is "...a mutually reinforcing system. The effect of one determinant is contigent on the state of others. Favorable demand conditions, for example, will not lead to competitive advantage unless the state of rivalry is sufficient to cause firms to respond to them. Advantages in one determinant can also create or upgrade advantages in others." (Porter, 1990, p. 72).

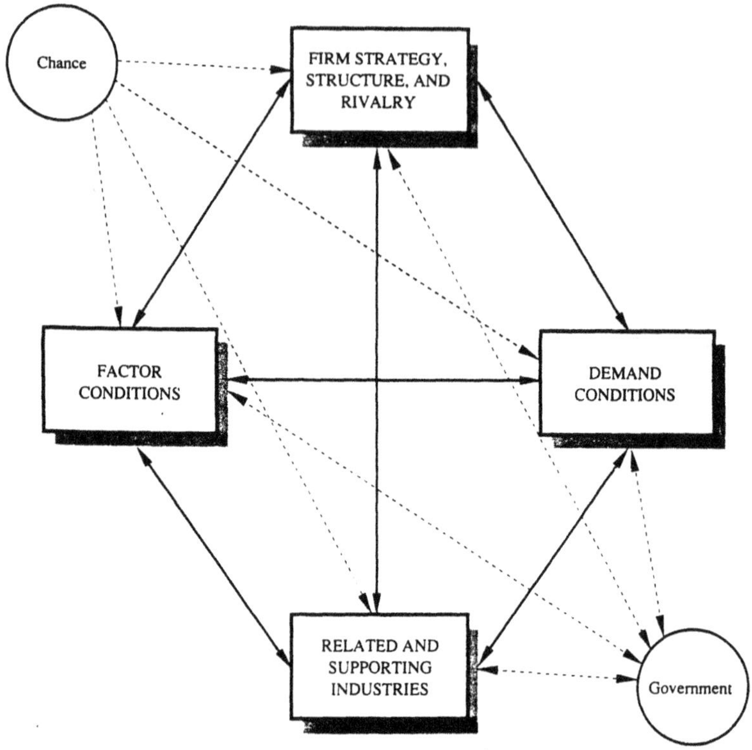

Fig. 1: Porter's "diamond" of "The Determinants of National Advantage" (Porter's 1990, p. 72)

Let us for a moment try to evaluate the different culture approaches mentioned above in relation to Porter's diamond. The approach of "corporate culture" is focusing mainly on internal attributes of single firms. In fact it is restricted to one of Porter's four boxes ("Firm strategy, structure and rivalry"). On the other hand Berger's term of "economic culture" is too general to be put into Porter's diamond, lacking mediating forces between the macro- and micro level of culture analysis. It is precisely such kinds of mediating forces Wilk's concept of industrial culture on Porter's "diamond" is supposed to explain though, from different scientific viewpoints. The problem of Porter's approach is, that it mainly focuses on the "rate" and "speed" of development rather than with the "content" of factors like technology and skills. Based on the economic tradition of abstracting from concrete analysis of technology and skills Porter's approach is insufficient regarding the more precise understanding of dynamic processes of competitiveness. Furthermore, Porter's model is based on a system-oriented functionalistic paradigm without really digging in the internal contradictions and the variously dialectical relationships between the institutions of the different industrial cultures.

The economic sciences have rediscovered the theme of culture. With the help of a concept that is designated as economic culture by various authors, the differences between various economic cultures have been studied and compared to macroeconomic factors, such as gross national product or, generally, the performance of national or regional economies. The point of contact between economic studies on the connection between culture and economy and our own studies on the topic of culture and production is the interest in a deeper understanding of what makes up the culture of a country or of a region with regard to the aspect of quality of the social organization of innovation. Here we assume that technology and economy are different forms of expression of this social process. This is where the interest in the rediscovery of the concept of culture in the social sciences stems from in the end. In this connection S. Wilks formulates his research and cognitive interest as an economist when he uses the term "industrial culture". "Using 'industrial culture' in this fashion requires that the origins, form and effect of culture be identified as part of a process of innovation" (Wilks, 1990, p. 147). He determines that it is not enough to attain a deeper understanding of the social organization of innovation as such, but rather that the concept of industrial culture is an excellent instrument for conducting cross-national comparisons.

2 Industrial culture approaches in this book

In this book we prefer the term "industrial culture". Compared to "corporate culture" industrial culture is not restricted to internal attributes of the single firm but includes the context variables as well. In this respect it is similar to Berger's concept of "economic culture" or Porter's "diamond". Nevertheless, we find an important difference between our approach and the approach of Porter. We focuse not only on the degree of technical differences and the levels of skills, but also on the content of technique and skills: which kind of industrial cultural forces are related to which kind of technologies and skills and how is this relationship influencing the market?

Despite Porter's theory of competitive advantage, which goes far beyond traditional economy, and his holistic approach, his economic theory has a basic difficulty in reaching the level of concrete technologies and skills. In the end, however, we have to reach this level of concrete content of work, technology and education as well as other factors of innovation if we wish to explain why, for example, a company stops buying its machine tools from a certain U.S. company at a certain point in time, but rather from a European or Japanese firm.

Still impressed by the realization that special national features in the social organization of industrial societies can obviously have far-reaching effects on industrial development and global market events, we have initially neglected the side of macroeconomic forces of technological development which has been touted by mainstream sociologists and economists, with a certain degree of one-sidedness, as a comprehensive explanatory approach.

The empirical orientation of industrial sociology in its research has been concerned with the conditions and consequences of industrial development in clearly defined and reasonably coherent economic and industrial regions - just for reasons of achieving the internal validity of its research results striven for. The differences in social development between regions and industrial countries which have, nevertheless, been revealed were, for a long time, interpreted as deviations and as an expression of nonsimultaneity in the historical process. Both express the fact that scientific thinking is deeply rooted in categories of natural science laws and of technological and economic determinism. This widespread narrowing of sociological and economic science is astonishing, first of all, because Max Weber clearly worked out the significance in his religious philosophy and in the development of modern capitalism. In dealing with this question, Clegg and Redding refer to a paradox formulated by Weber himself which in its short reception shows a culturalist as well as a globalist view of social development processes. "For Weber, the existence of culture and meaning in the economic code of modern capitalism would have been an agreeable paradox. Agreeable, because his basic conception of economics was not one which stressed it as natural science, but instead regarded it as a cultural science. The paradox would arise because it was Weber's view that although modern industrial capitalism had been forged in the heat of religious values and culture, it was set in a mould from which these sources of meaning were draining away. The pan-cultural value of "rational action" would transform the contours of modern capitalism to a uniformity in which cultural value was absent." (Clegg and Redding, 1990, p. 51)

The paradox is, therefore, that, with the help of the category of culture, Weber describes the structure of the capitalist society, which, though strongly influenced by religion and culture in its historical development, can do without culture as an explanatory category in the end. Weber's future view of an iron cage of bureaucracy sees man as an insignificant cog in an all-infesting machinery. This picture of industrial society is increasingly asserting itself in sociology - the differences between industrial cultures disappear from view. Herbert Marcuse's "One-Dimensional Man" and finally Braverman's "The Degradation of Work in the Twentieth Century" mark a debate on capitalism in which culture has definitively vanished as an aspect of diverging social developments.

In this respect we agree with Berger's approach "...to avoid the distortions of so-called convergence theory, which trivializes very important economic, social and political differences." (Berger, 1987). But we do not accept Berger's system approach based on a functionalist paradigm. We prefere a dialectical approach focusing on the oppositions of different institutional forces behind innovations and the dialectical interaction between convergent and divergent forces of technology development.

Nevertheless, even in this book no uniform concept of industrial culture is presented. In fact, the authors of this book disagree to a certain extent regarding how the dialectical forces behind technological development should be termed. Rauner/Ruth focus on the relationship between "culture" and "society". In their view culture and society are two different ways of looking at a social system, they stand in opposition to one another. Society is usually taken to include the structural characteristics of a social system, such as power, rule, class, equalities and inequalities. Up to now sociology has primarily concentrated on these structural aspects of society. Culture, on the other hand, consists of symbolic representations, such as norms and values, religion and law, ideology, concepts, etc. The German Conference of Sociologists in 1988 undertook a commendable attempt to designate one-sided approaches in the view of the interrelation between culture and society and to emphasize the conflicting interrelation between structural (social) and cultural forces in the process of social change.

Rauner/Ruth's concept is based on an understanding of the relation between structural and cultural elements as a dialectic one. Like R. Münch, they assume "that social analysis remains blind, i.e. one-dimensional, superficial, over-generalized, simplified and speechless, if it doesn't relate the dimension of social reproduction, which it emphasizes, to that of cultural reproduction. Cultural analysis exposes deeper layers and tells us more about the meaning of social development, but it remains vacuous, - if it isn't related to the various dimensions of social reproduction." (R. Münch, 1989, p. 702)

Following this view, Rauner/Ruth interpret technological development as a cyclical process, in which the divergent elements (influenced by industrial culture) and the convergent elements (influenced by industrial society) are dialectically related to each other and always take effect simultaneously. At an initial stage the cultural elements are dominant, while later the socio-industrial forces become more effective and invert the initial divergent phase of technological development to a convergent one. At the same time, industrial culture exerts just as much influence on the creation of industrial society as the other way around.

Rasmussen's interpretation of the dialectical process behind technological development is slightly different. In his view the distinction between 'society' and 'culture' is rooted in a dualistic tradition that presupposes a particular domain for 'culture' as well as for 'society' (or 'social structure'). In opposition to this dualistic tradition he finds it quite impossible to locate culture in a particular subjected area. There is no culture-free domain. Culture is inherent in social, economic as well as technical 'forces' or 'variables'. According to Rasmussen the term 'culture' should not be restricted to the world of 'symbols' or 'values' as the dualistic thinking presupposes. In his view culture is interpretation and interaction instead of interpretation of interaction. Consequently, the dialectical movement is not a process between 'culture' and 'society' or between 'culture' and 'economy'. The movement is internal in the particular industrial culture itself as well as between different industrial cultures.

This disagreement on the general conceptualization of culture does not prevent a very similar perspective regarding the more empirical oriented levels of analytical approach. Common to all approaches of this book is the looking for mediating 'actors' between the macro-oriented concept of 'economic culture' and the micro-oriented concept of 'organizational culture'. Action-oriented perspectives of industrial culture are in focus in nearly all the articles of this book. Thus Klaus Ruth discusses 'action - the missing link' and develops a model of a 'socio-technical Action system'. Lauge Rasmussen discusses industrial cultures as 'multi-dimensional institutional settings as well as actors of transcending and maintenance of these settings'. Finally, Martin Corbett, Jürgen Friedrich and Thomas Binder/Klaus T. Nielsen focus on the 'designer/user' and 'producer/user' interrelationships as a dynamic perspective of new industrial design cultures.

The concepts of industrial cultures possess the potential to widen the scope of understanding competitiveness and technological development. Global market trends make it still more obvious, that forms of competition are gaining in significance in which pure production factor explanations are becoming more and more insufficient. More and more it becomes clear that industrial cultures form a necessary background as well as process oriented pattern of actions in which convergent and divergent forces interact and dominate variously in different periods of the technological development cycles.

Consequently, it might seem too 'modest' just to consider industrial culture as a "useful corrective to the oversimplifications inherent in the determinist-strategic choice debate" (Martin Corbett, chap. 3). Though we agree on the critique of technological determinism and other forms of determinism too, the industrial culture apporach is <u>much more</u> than a 'useful corrective'. It is a complete different way of interpretation and understanding competitiveness and technological development. This is stressed by Klaus Ruth, in particular, when he writes that "... the fundamental axiom of the industrial culture approach assumes the existence of many industrial cultures, each of which has its own strengths and weaknesses as well as its own dynamics of development". (Klaus Ruth, chap. 6).

Though a <u>widened scope</u> of understanding competitiveness and technological development is useful from a scientific as well as from a pragmatic industrialist' viewpoint, it raises a number of new <u>epistemological</u> problems too. How can such all-embracing approaches be fulfilled on the empirical levels? How do we select 'important dimensions' of the 'entire' industrial culture or cultures to be analysed, evaluated and mutually interrelated?

Several of the approaches of this book prefer a <u>dialectical</u> understanding from a more 'neutral' system approach. The system approach of 'mutually interrelated variables' at least overcomes 'one-way' determinism. Nevertheless, the question is if such a 'two-way' relationship is too weak a model for interpretating technological development? Contrary to system approaches, a dialectical approach of industrial culture focuses not only on mutual interrelationships as such, but on <u>oppositional</u> interrelationships. Examples of such oppositions are mentioned in several articles in this book. Rauner/Ruth focused on the 'bipolar reciprocal interrelations' between 'social institutions', 'educational institutions', 'industrial organization', 'government policy' and 'psychology'. Rasmussen mentions the opposition between 'planning and executive tasks', between 'educational traditions and development of job structure' and between 'producers and users of production technology'.

Several other articles mention the opposition between the 'planning and execution', the 'hard' system design thinking and the user needs in practice. On the empirical level of comparative studies of machine tool industry in different industrial cultures, Gabrielle Laske concludes that "... the German engineering philosophy of only creating ambitious and advanced machine tools concept is starting to become a hindrance for the whole industry. To follow this track means to lose extensive market shares ... of low-priced, easy-to-operate machines". (Laske, chap. 9). In opposition, the 'analytical' approach of German designers, Eckehard F. Moritz found a 'synthetic' approach among Japanese designers of machine tools and a 'pragmatic' approach among American machine tool designers. Though such kind of culture-bound distinctions may seem a bit too simple, it nevertheless contends important oppositions internal in the different industrial cultures worth considering more in detail.

Thus the articles in this book provide an opening towards a more comprehensive understanding of competitiveness. Based on empirical surveys of the machine tool industry in different countries and on an extensive theoretical and methodological discussion on the concept and analytical dimensions of industrial culture, this book provides a promising 'agenda' of understanding competitiveness as a dynamic interpretation and interaction between institutional oppositions as well as action oriented patterns of design philosophy and user priorities. Far from presenting all the 'answers' to this widened approach of analysing competitiveness, it rather begins directing research in more constructive ways of understanding how the seeds of the future are enfolded in the oppositions shaping the present industrial cultures.

REFERENCES

Barney, Jay (1985): *Dimensions of informal Social Network Structure*, Social Network 77:1-46.

Berger, P.L. (1987): *The capitalist revolution*, London: Wildwood House.

Clegg, S.R., S.G. Redding (Eds.) (1990): *Capitalism in Contrasting Cultures,* Berlin, Walter de Gruyter & Co.

Dyson, K. and Wilks, S. (1983) (eds): *Industrial crisis*: a comparative study of the state and industry, 26-66, Oxford: Martin Robertson.

Hofstede, G. (1980): *Culture's consequences*, London: Sage.

Morgan, Garett (1986): *Images of Organization*, London: Sage.

Münch, R. (1989). Gesellschaftsanalyse und Kulturdeutung. In: Haller, Hoffmann-Nowotny, Zapf (eds.). *Kultur und Gesellschaft.*

Peters, T.J. and R.H. Waterman (1982): *In search of excellence*: lessons from America's best-run companies. New York: Harper and Row.

Porter, M. (1990): *The Determinants of National Advantage*, N.Y.: Free Press.

Wilks, Stephen: *The embodiment of Industrial Culture in Bureaucracy and Management* in Clegg, S.R. et al (1990): *Capitalism in Contrasting Cultures*, New York, 1990.

Section I

Scientific Concepts of Industrial Culture

Industrial Culture and Production - Towards a New Research Approach

by Felix Rauner and Klaus Ruth

1 Introduction

The elaboration of a research approach in *industrial culture and production* could be misinterpreted as opportunistically fawning to a trend which is attempting to cover up the political and economic **structures** of industrial production with the term *culture*. Certainly, the fashionable and strategic use of the terms *organization culture* and *industrial culture*, especially in the field of management sciences, and the way they are used in the revival of the socio-cultural debate, need to undergo an ideologically critical analysis (Münch 1986).

Industrial culture signifies a research approach not aimed at reinterpreting the conditions of industrial and capitalist production as a form of *culture* - a neutral and exalted sphere of human existence in which the principle of criticism is ideologically substituted by that of contemplativeness.

Our interest in the elements of industrial culture which influence the development and application of technology is the scientific result of the reception of techno-sociological and techno-historical research which has reached the conclusion that an historical process is not a continuous and linearly cumulative development of social experience and the forces of production, expressed in turn in the deterministic picture of technological and scientific progress.

It has only been in the course of the last decade that social scientific research into technology has turned to the shape of technology as an expression of industrial culture.

This fundamental change leading to an understanding of technological development as the result of the interplay of work and technology was triggered by Kuby's (1980)[1]) critique of determinism. He has used the example of the spinning machine to refute that techno-historical paradigm which claims that technology advances through steady accumulation of social experience from the craft tool to the self-acting machine. Likewise, Howard Rosenbrock (1984), David Noble (1986) and Hans-Dieter Hellige's (1984) fundamental considerations and investigations of technological development have contributed to the recognition that the shaping of technology and the latter's application in the work process oscillate between industrial-culturally divergent and convergent development stages.

The two concepts of Technology Assessment (TA) and of the Humanization of Working Life (in German: "Humanisiserung des Arbeitslebens", HdA), which are inherently preventive and reactive rather than shaping-oriented, were given a new direction by the change from technological determinism to the shaping of technology. This shift has taken shape as the research into work and technology (Rauner, 1989).

Our definitions of *society* and *culture* in the concept of *Culture and Production* are not interpreted as concepts which exclude each other, but as dialectically related opposites which are essential for each other's existence. In the history of social sciences it is the structural elements, especially the macro-economic forces and their development, which are recognised to have influenced everything else. Culture then appears to be an aspect of the superstructure and represents the systems of norms and values and their institutions, as well as the world of values, which is beyond that of functional reasoning.

From the socio-cultural point of view, the structural elements of society appear to be cultural products, or they disappear completely from view. From the point of view of industrial culture, the uniqueness and peculiarities of a society are emphasized. Common features shared by certain industrial cultures are regarded as general features and are thus culturally determined.

In contrast, the *main stream* sociological perspective views world society as a whole. Consequently, exceptional features of certain industrial cultures are interpreted as deviations from the basic model of industrial society or from the structures of industrial societies. From this point of view, deviations can in the end only be

---

[1]) Unlike Karl Marx, Thomas Kuby does not regard the 'Spinning Jenny' as the first machine to arise from a qualitative change at the height of craft development, which is seen as merely a transitory stage in the continuously ascending development of technology. Rather Kuby views the 'Jenny'as craft's final response to the management problems of expanding British capitalism, a problem alien to the crafts. "Arkwright's spinning machine cannot be accounted for by the current paradigm of the historiography of technology. The transition from the craft tool to the machine did not occur via the 'Jenny'as an intermediate stage, was not a succession of relatively small steps. Instead, this transition was a tremendous leap between two different systems of productivity as well as between technical means and work of a completely *different nature* (not only state of development)" (Kuby, 1980, 88, transl. by the author).

explained by geographical hindrances, without losing sight of the quality of the structures of the industrial society.

2 Industrial Culture and Industrial Society

An antithetical use of *industrial culture* and *industrial society* is in a way a continuation of the well-known debate on capitalism versus industrial society which was initiated by Adorno and others (Adorno 1974) and later carried out by certain branches of industrial sociology until the late 1970s. The sociological debate on capitalism and industrial society - built up on the works of Aron (1974) on industrial society - was also an implicite (and occasionally very explicite) discussion of the categories class and social strata, pauperization and affluent society. Thus the terms capitalism and industrial society indeed have different political implications but in our view they still share the same epistemological paradigm, i. e. both concepts aim at common structural determinants of industrial-capitalist development, whereas the debate on industrial culture on one side and industrial society and even post-industrial society in Bells sense (Bell 1973) on the other indicates a discussion between two concepts originating in different paradigmatic spheres.

Applied to technological developments the concept of industrial culture is intended particularly to convey the insufficiency of any approach that identifies the level of technological development reached by a specific historical period for use as a yardstick against which specific industrial countries are measured. Rather, this concept denotes the fact that comparable industrial countries have indeed produced different qualities of technology for identical or comparable purposes. What is true for the development of technology equally applies to the latter's application: this often varies with industrial culture.

The term *industrial culture* - in preference to *industrial society* - helps systematize the peculiarities that characteristically distinguish industrial countries as regards the development and application of technology. The term moreover helps to compress these peculiarities into a theoretical concept so as to help stimulate further-reaching research hypotheses and interpret the results of empirical research. In the theory of industrial society, differences that become apparent in a comparison between industrial countries and regions represent the "flawed surface", as it were, of historical everyday life, beneath which the real basic structures of industrial society in general have to be unearthed. This theory thus aims to reveal the laws of locomotion of capitalist society and hence of industrial technology. Braverman's *Labor and Monopoly Capital* (1974) is merely one example typifying an era of research into industrial relations in which the differences in the social development of industrial countries had about the same relevance as the disturbing effects of real nature had for the sciences. These aimed to discover the "real laws" of nature via a painstaking process of controlling conditions.

The concept of industrial culture, by contrast, not only emphasizes the differences between comparable industrial societies, but even elevates these to the level of characteristics. The different paths pursued with regard to energy technology, for instance, by the US, Sweden and the FRG, thus do not represent different forms of the same historically given, technologically superior energy technology predominating over the differences discernible in a comparison of these countries. Rather, the different actual technologies are expressions of quite different industrial cultures. Industrial-cultural determinants of the development and application of technology thus correspond precisely with those social conditions that are not common to all industrial societies, but which mark their **characteristic** differences.

It was quite a while ago that the first attempt was made to account for *Japanese Technology* (Moritani, 1982) not merely by recourse to economic categories and theories: "Comparative technology basically means looking at the technology of different nations in relation to their climate for technology. In its broader sense, it is the study of technology through cultural and historical tradition" (Moritani, 1982). In order to understand the Japanese superiority in the field of modern technologies, Moritani claims, one has to spend considerable time unearthing the cultural, social, and historical roots of technological development. "In other words, the culture and customs of various countries are manifested more and more clearly in the products they manufacture" (Moritani, 1982, 21). Moritani identifies one of these roots as "..the Japanese traditional craftmanship as seen in bonsai and netsuke, the painstaking creation of delicate miniature masterpieces ". This tradition has persisted into the present and can be found "...in Japanese compact and high performance precision industrial products" (Moritani, 1982, 31).

Our concept of industrial culture is based on an understanding of the relation between structural and cultural elements as a dialectic one. Like R. Münch, we assume "that social analysis remains blind, i.e. one-dimensional, superficial, overgeneralized, simplified and speechless, if it doesn't relate the dimension of social reproduction, which it emphasizes, to that of cultural reproduction. Cultural analysis exposes deeper layers and tells us more about the meaning of social development, but it remains vacuous, ..., if it isn't related to the various dimensions of social reproduction" (R. Münch, 1989, p. 702).

The question is - how can a dialectic understanding of the combination of cultural and social determinants be used beneficially in the analysis of the development and application of technology?

The empirical research into technology envisaged by the current *industrial culture* research projects is aimed at identifying the balance, or the temporary imbalance, between the structural and cultural elements which determine the development and the application of programmable automation (PA) over a period of time. By the determinants of industrial cultures we mean those elements of technological development and application which lead to *diverging* technological methods.

As opposed to the determinants of industrial cultures, we categorize as elements which determine the development and application of technology in industrial societies those global and international structural forces which have a *convergent* effect.

In each technological sector, the **diverging forces** caused by the different industrial cultures, and the structurally determined **converging forces** take effect at the same time - albeit in a cyclical relationship. On the basis of this dialectic mental picture, we arrive at the hypothesis that a new form of technology, in the phase from its creation to its introduction on the market, is strongly determined by industrial culture

- not forgetting that it is the structure of the industrial society concerned which form the basis for any technological development.

During their development phase, however, new techniques are greatly influenced by regional and cultural factors. Techniques developed at the same time, but in different places or in different countries, are, in their initial phase, not so much subject to the structural forces. These then take effect, in the case of industrial technology, at the latest on its introduction onto the market. As technical products are marketed and specific market sectors are internationalized, the structural forces with convergent effects increase. At first, market cleansing, license markets, national and international standardization policies and the international character of science restrict the divergent technological developments. Finally, technological development narrows down under the increasing pressure of the selective forces of industrial societies.

Thus, technological development can be interpreted as a **cyclical process**, in which the divergent elements (influenced by industrial culture) and the convergent elements (influenced by industrial society) are dialectically related to each other and always take effect simultaneously. At an initial stage the cultural elements are dominant, while later the socio-industrial forces become more effective and invert the initial divergent phase of technological development to a convergent one. At the same time, industrial culture exerts just as much influence on the creation of industrial society as the other way around.

An exemplary convergence/divergence cycle occured with the CNC developments: In the beginning period of CNC history many different concepts have been developed (external programming; data input via tapes and optimizing by direct input; multi processor controls; manual data input at the machine supported by programming aids, later supported by menu graphics; shop-floor oriented programming systems) but the middle of the 1980s was the turning point after which the converging forces got more influential, i.e. the internationalization and globalization of production overruled the presumably industrial culturally determined diverging forces. As examples might serve the harmonization of the human/machine interfaces, the standardization efforts (e.g. the ISO codes, etc.). Generally the converging forces exert a strong impact on the system cores of CNCs but beneath these convergencies there are shaping scopes which allow the user-specific generation of the work-relevant layers of systems and thus diverging tendencies below the converging main stream.

3 Some Theoretical and Terminological Remarks on Industrial Culture

The necessity to develop a new research approach in technological development and application has to be seen as a consequence of the obvious deficiencies of other theories in that area, such as purely economic approaches in the sphere of application, *demand-pull* and *supply push* approaches in the field of technological development (Dierkes 1989, 158). It was the critical discussion of the mentioned theories which led to the claim to include socio-political elements such as the general techno-political conditions, the dominating orientations in the (techno-) scientific system, the anticipation of future application potentials, the specifities of technology developping organizations, etc into an appropriate theory of technology development and use. The industrial culture approach contains all these elements and furthermore includes genuine cultural variables. Besides the enlarged contents of the term it is a new epistemological aim which is followed up by the industrial culture approach, i. e. to give reasons for different developmental and application paths in the field of programmable automation technology. It is on the one hand the demand to explain national differences but on the other hand it is the aim to explain converging tendencies and to elucidate the relationship and the mutual dependency of both.

Evidently a term like industrial culture which is a combination of two different terms that belong to originally different fields of application has to make its roots and influences explicite - especially if the term is supposed to be the central category of a research approach. The fact that the attempts to define the term refer to different fields of influence does not mean that industrial culture itself is a combined and inconsistent residual category.

Any attempts to understand the new leading part of culture within many scientific disciplines and particularly in organization sciences, some sub-branches of industrial sociology etc. obviously depend on the particular research tradition and the personal point of view. In the widely prefered approach of cognitive anthropology culture, organization culture, and thus industrial culture is grasped as a set of shared knowledge that accompanies and influences the behaviours and practices of a certain group (sub-group or nationality). To this narrow cognitivist approach we would add, that besides shared knowledge also common values, symbols, habitual ways of thinking, man ifestations of unconscious dynamics and (colective and individual) dreams of technical images (be it daydreams or nightdreams) have to be included (Dahmer 1985). Following M. Weber (1976), culture (and thus industrial culture) can be seen as a stock of shared values and implicit rules out of which individuals obtain their patterns of interpretation and action.) In addition to the Weberian approach in which the crucial element is the *meaning* of technical artifacts that is ascribed to them by interpretation of human beings we would emphasize a social process of communicative acts in which the persons involved make sure of the meaning of technical artifacts, work, etc (Habermas 1981). In this approach culture is recognized as a kind of social discourse that is centrally based on a perpetual process of interpreting - Habermas calls it *interpreting acts*. In this sense culture contributes to the social construction of reality (Berger/Luckmann 1966) and furthermore develops an efficacy in the real life of construction (of technical artifacts).

In a functionalist view culture regulates the purposes/means relation that underlies collective and individual patterns of action (like design and application of technology)[2]. The relevance of the functionalist view for the industrial culture approach is

---

[2] This approach has to be recognized as an enlargement of the reductionist approach to technology/technique which defines it simply as a purpose/means relation.

pointed out below. In a beforehand step we will develop the substantial elements of the industrial cultural conception.

Before discussing the importance of the structural interplay of some industrial cultural key dimensions and their significance in an approach that aims at elucidating the influential variables of differing technological developments as well as different conditions of technology application we will present some comprehensive key dimensions of industrial culture. This dimensions link the abstract category *industrial culture* with the operationalized items of empirical investigations. The industrial cultural dimensions thus denote fields which feed the theoretical category. One might say that in this respect it is a mixed combination of different fragments originating from different scientific disciplines, but nevertheless industrial culture is a consistent central term in respect of its interpretative potential.

4 Dimensions of Industrial Culture

In the industrial-cultural approach we assume, that especially the processes of design and use of technology depend upon and are intermingled with extra-technological parameters. Apart from socio-economic and institutional variables, we believe industrial culture to exert an important influence on technology design and application. The concept of industrial culture is based on a set of categories, highly interconnected and mutually influential dimensions and variables which form an interpretative frame. To the most important categorial structural dimensions we count:

1. Social Institutions
2. Industrial Organization
3. Educational Institutions
4. Government Policy
5. Psychology

4.1 Social Institutions

Among the most important dimensions of *social institutions* we count a stock of cultural values, which influence both design and use of technology. An important variable in this field that contributes to national traditions is the attitude towards technique and technological progress which in some aspects differs considerably in West Germany and Japan: Over the past few years many opinion polls detected a critical way of thinking about technology within the German youth, whereas the Japanese youth seems to be slightly more optimistic about the future problem solving capacity of technological developments (Jaufmann, et al. 1989 and Moritani 1982). According to our hypothesis regarding design and development of technology a critical attitude - or at least an ambivalent non-optimistic view of technology, its usefulness, its consequences, etc. - seems to be a prerequisite for a subjective awareness of the social and work oriented shapability of technology. Engineers who do not any more believe in a deterministic objectivity of technology design - surveys of German engineers and engineering students indicate this as a growing attitude among this group (VDI-Nachrichten 1989) - will probably search other (new) ways in machine design in contrast to their colleagues who are not so much convinced of the existence of broader scopes of design alternatives.

At the *application side* an ambivalent view of technology will possibly transform itself into a demanding attitude towards technology and work (the contents, the organization, the self-determining capabilities, etc.) among (skilled) workers. (See the following paragraphs.) Both tendencies converge and form an element of tradition which itself is recognized as a key dimension of industrial culture.

In our view it is the social assessment of work, the work-leisure proportion and a historically grown demanding attitude towards the contents of work that affect the industrial-cultural sphere of technological design and use. Especially the German machine tool sector with its highly skilled labour force is caracterized by a dominating demanding attitude of the workforce towards the work contents. In addition to the activities of the works concils the skilled machine workers individually or in informal groups always tried to improve the work style, the work contents, etc (-Kern/Schumann 1986). Their pretentious demands towards work constitute very obviously a national tradition and thus an element of German industrial culture that is complemented by a pronounced work-leisure differentiation. The Japanese case displays at least regarding the last feature some considerable differences (Weber 1989).

Furthermore we see the techno-scientific problem-solving perspectives of engineers as a very important industrial-cultural reference size in the technology design process. These problem-solving perspectives represent a historico-socially determined limitation of the entire horizon to which the knowledge and experience of a given technological branch extend. As a result, only selected elements out of the total amount of feasible technical solutions, objects and methods are perceived as viable options and are elaborated upon. Thus the problem-solving perspectives are embedded within nataional traditions and exercise a strong impact on design and use of technical artifacts (Hellige 1984).

To those social institutions which are of particular relevance in the German machine tool sector we count the manufacturer/customer relationship. The formal conditions of ordering and delivery, the joint definition of requirement profiles as well as informal manufacturer/customer relations contribute to an industrial cultural tradition which allows the customers to have some influences on the shape of the machines delivered. We expect national variances in the mentioned tradition which is probably of great significance. We recognise this as an important factor to be included into a comparative investigation in industrial culture.

The industrial relations and the institutionalized patterns of capital/labour relations in Germany can be roughly circumscribed as a partnership of the two sides of industry. A cooperative mode of conflict solving is dominating in the shape of an institutionalized bargaining of unions and shop floor representatives. A peculiarity of the German machine tool sector is the representation of particular interests regarding working conditions, socio-technical shaping potentials, wage fixing, etc. This direct and informal interference of production workers is owed to their relatively strong position in the production process and is resulting in pragmatic compromises (Kern, Schumann 1986). Other countries, such as the Scandinavian, indicate varying forms of industrial relations and collective bargaining which both constitute a differing industrial culture.

As a systematic completion we have to value the significant impacts of the *social contract at company level* on the industrial cultural sphere. This conception assumes that any company is not only determined by economic, technological and administrative prerogatives but also by human factors. In this view companies are preferably seen as social systems which function along certain "rules". The social contract at company level is based upon the consensual and mutual integration of all groups within the companies. It comprises a common understanding of a system of informal rights and obligations concerning policy of employment securing, continual training courses at company level, wage fixing and control of performance, etc (Hildebrandt, Seltz 1989). While the social contract at company level takes different shape in different countries it contributes to different industrial cultures.

4.2 Industrial Organization

Different national patterns of *industrial organization* which cause either a more Tayloristic, a craft-oriented or a more flat hierarchical organization of work contribute to an industrial-cultural approach to technology. The dominating organizational orientations of a country can appropriately been understood only in terms of an economic- and social-historical approach. In the German case a social-organizational corporativism, which had its roots in the medieval guilds was taken over into the craftsman's shops and later on influenced the organizational structure in many industrial branches such as the machine tool sector, metal manufacture, etc. In this view the German "Meister/Facharbeiter" system on the shop floor level is concordant to work/craft oriented, hierarchically flat organization patterns. Despite of many similarities compared to the German example the British case (Sorge 1985) and to an even larger degree the U.S. seem to differ in the historical formation of current organizational work structures which in the U.S. through many years have been largely oriented towards Taylorism (Braverman 1977).

4.3 Educational Institutions

In our concept we believe *educational institutions* to be of extraordinary relevance for national industrial cultures and thus for the design and use of thechology. In particular we see the educational systems, the contents of educational processes, the relevance of skilled work and the patterns of skill formation processes as the powerful key variables in this category. The historical genesis of the educational system in Germany is distinguished by government-sponsored education and vocational training. This policy formed the base of the dual vocational training system - qualification **at** work and **for** work - with its institutionalized combination of theory and practice. This resulted in a relatively highly qualified skilled labour.

The underlying cultural orientation is the so-called skilled-labour-demand approach in which the educational system provides skills, qualifications and knowledges in sufficient quantity and quality. Education and the vocational system is forced upon the individuals whereas other cultural contexts shape a human capital approach in which individuals decide on their personal investment in qualification and education (Lenhardt 1987).

In the course of history the institutionalized skilled-labour-demand system caused a large amount of skilled labour (Facharbeiter) which is one of todays characteristics of the German economy in general and of the machine tool sector in particular. The mutual dependencies of this circumstances with other above mentioned industrial cultural variables is evident.

4.4 Government Policy

Government policy is very influential to industrial culture and in our view the different policy fields are largely determined by industrial relations (capital/labour relations) and the social contract at company level (see above).

To exemplify the significance of government policy we refer to the German government-sponsored WOP-programme on "shopfloor-oriented programming of CNC-machine tools" (Nuber, et al., 1989; WOP Conference Materials, 1987 and the contribution of Brödner in chapter 4 of this volume). During the last years this programme was carried out by a consortium of machine tool producers, machine tool user companies and research institutes with the goal to develop user-oriented programming techniques which are highly adapted to the specific productive processes and to the skills of the "Facharbeiter". As a result the developed programming system is used far beyond the involved user companies. Many machine tool producers that have not participated in the WOP-programme are now producing CNC-machine tools with shopfloor-oriented programming facilities. This development is still a strong movement in the German machine tool building. This sensible government policy gave highly efficient inputs and caused strong spin offs. The whole WOP programme of course was enabled by an industrial culturally determined consensus of all relevant social groups and by the insight into the necessity of governmental support in the fields outside of basic technologies. Some strong similarities regarding state policy displays the Japanese MITI which in this respect must be tied up in a similar industrial cultural field.

4.5 Psychology

In addition to the above mentioned categories the industrial culture conception includes *psychological* elements such as national socialization types. Particularly the different cultural patterns of constituting subjectivity (especially the role of vocational socialization) and the importance of individuality within the individual/society-relation need to be considered in a comparative industrial cultural approach (Vogel 1983).

To the important feature which is located both in social and individual psychological sphere we count *dreams* (colective and individual dreams) and images of technique - which are the emotional/intuitive complementary variables to the above mentioned problem-solving perspectives. On the level of human civilization Mumford (1977) already pointed out the importance of dreaming in the process of mechanical inventions. If we project this important insight to the level of culture (industrial culture respectively) there is some considerable evidence that the engineer's anticipatory day- or nightdreams of technical artefacts will be a) influential for their machine design and b) different within differing cultural systems (Dahmer 1985). In our view these dreams are intermingled with certain images of techniques (Gloede, et al. 1989) and both are taking effect on the design processes and products - so our hypothesis.

The intensity of the impact of all these factors on design and use of technology varies with a country or culture. In the following we want to sketch the structural influence of the above mentioned industrial cultural dimensions.

5 The Structural Interaction of Industrial Cultural Dimensions

A view at the structural interdependence of the industrial cultural dimensions allows to assess their effects on technological systems design and use. The below illustrated figure on the *Sphere of Industrial-Cultural influences* indicates a so-to-say "gravitational field" of industrial-cultural influences on the design and application of technology. The arrows between the five corners of the figure always aim in both directions, this indicates that it has to be imagined as bipolar *reciprocal interrelations*: All dimensions are connected with each other and all are influencing each other mutually. Evidently this picture is not a static model but a snapshot of a dynamic interrelation.

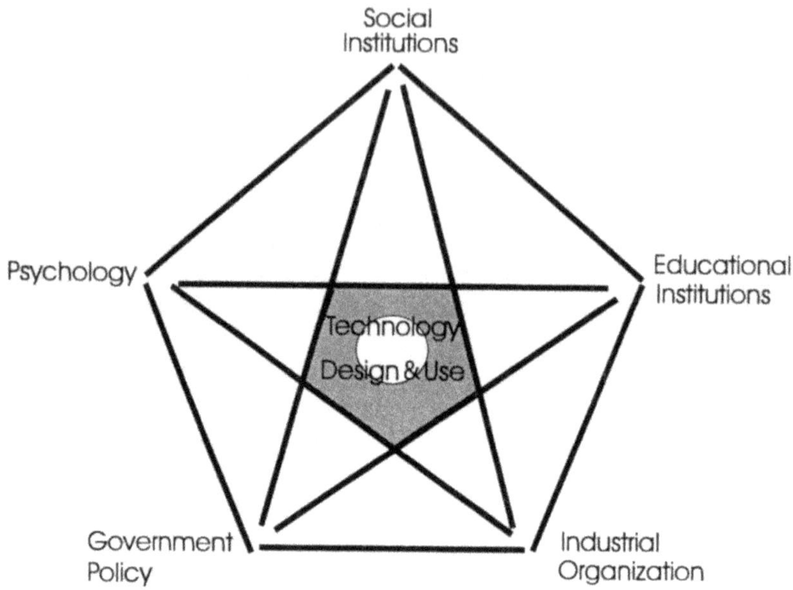

Figure 1: The Sphere of Industrial-Cultural Influence

Within the influential field exists an inner core consisting of techno-logic design and use parameters which are of course not neglectable. The techno-logic issues itself are intermingled with non-technical, economical, social and cultural influences and thus need to be analytically seperated from these - which seems to be a difficult task because there is a kind of grey area around the techno-logical core of design and use which is recognized as appertaining to the inner core. As difficult as it may be to detect non-technical influences in this grey area we believe there exist some "surrounding layers" where it is really evident that non-technical in our approach industrial-cultural - variables determine the possibilities of technology shaping and application. The shape of human/machine interfaces in the CNC-machine tool domain might serve as a short example: In a certain period of CNC development (around the end of the 1970s) the shop-floor oriented programming features (colour graphics display, dialogue-oriented programming languages, etc) have with some exceptions not been realized because this orientation lay outside the considerations of the pure techno-logic on the one hand. On the other hand the development of shop-floor oriented CNCs has not been pushed because it was not concordant with the industrial-cultural preconditions of the then dominating CNC-developing country, the US.

The strengthening of Japanese and German machine tool producers offered new shaping possibilities that were influenced by differing employment conditions (in small and medium enterprises with small batch sizes, etc) and industrial cultural factors.

The structural interplay of the industrial cultural dimensions, their specifically shaped sphere of influence, both are of particular interest in a comparative investigation in industrial cultural determinants of technology design and use.

6 Concluding Remarks

The sketched approach is reflected in various research projects. Most of the mentioned industrial cultural key variables were operationalized in several comparative research projects treating with CNC technology. As a result, our investigation no longer focuses exclusively on technological developments in machine-tool building as the unambiguously independent variable. Instead, we interpret technological developments (including those in production engineering) also as adaptations to factors or realities that are to do with social structures, organizational patterns and qualifications of labour - in one word: *industrial cultural factors*. The emphasis lies on the unity of both diametrically opposed and inconsistent analytical approaches, because each can claim a grain of truth. This is why a dialectical analysis of the relation of technological developments and industrial cultures is the most adequate: Industrial culture constitutes technology or relevant dimensions of the effect and development of technological innovation and conversely, technology has an impact on a given industrial culture.

REFERENCES

Adorno, Th. W. (1972): Spätkapitalismus oder Industriegesellschaft? In: Adorno *Soziologische Schriften I*. Frankfurt.

Aron, R. (1974): *Die industrielle Gesellschaft. 18 Vorlesungen*. Frankfurt (French original published in 1962).

Bell, D. (1973): *The Coming of Post-Industrial Society*. New York.

Berger, P. L. and Luckman, Th. (1966). *The Social Construction of Reality*. New York.

Braverman, H. (1974). *Labor and Monopoly Capital*. New York/London.

Dahmer, H. (1985). Surrealismus, Psychoanalyse, Politik. In: *Neue Rundschau* 2.1985.

Dierkes, M. (1989). Technikgenese als Gegenstand sozialwissenschaftlicher Forschung. In: Friedeburg andJacobi (eds) *Konzepte sozialwissenschaftlicher Technikforschung*.

Gloede, F. and H. Bücker-Gärtner (1989). Autonome Technik oder Technik als soziales Projekt - Technikbilder und Technikbewertung in der Bevölkerung. In: *Kultur und Gesellschaft*. Zürich.

Habermas, J. (1981). *Theorie des kommunikativen Handelns*. Frankfurt/Main.

Haller, M., Hoffmann-Nowotny, H.-J. and Zapf, W. (eds) (1989). *Kultur und Gesellschaft*. Frankfurt/New York.

Hellige, H.-D. (1984). Die gesellschaftlichen und historischen Grundlagen der Technikgestaltung als Gegenstand der Ingenieurausbildung. In: Troitzsch and König (eds) *Lernen aus der Technikgeschichte*. Düsseldorf.

Hildebrandt, E. and Seltz, R. (1989). *Wandel betrieblicher Sozialverfassung durch systemische Kontrolle*. Berlin.

Jaikumar, R. (1986). Postindustrial manufacturing, *Harvard Business Review*, 11/12, 69-76.

Jaufmann, D., Kistler, E. and G. Jänsch (1989). *Jugend und Technik*. Frankfurt.

Kern, M. and Schumann, M. (1986). *Das Ende der Arbeitsteilung*. München.

Kuby, Th. (1980). *Vom Handwerksinstrument zum Maschinensystem*. Berlin.

Lenhardt, G. (1987). Bilder und Zerrbilder von USA. In: *Leviathan*. 4/1987.

Lutz, B. (1989). Introduction to "Kultur und gesellschaftliche Entwicklung". In: Haller, Hoffmann-Nowotny, Zapf (eds). *Kultur und Gesellschaft*.

Moritani, M. (1982). *Japanese Technology*. Tokyo.

Münch, R. (1986). *Die Kultur der Moderne*. Frankfurt.

Münch, R. (1989). Gesellschaftsanalyse und Kulturdeutung. In: Haller, Hoffmann-Nowotny, Zapf (eds). *Kultur und Gesellschaft*.

Mumford, L. (1977). *Mythos der Maschine*. Frankfurt

Noble, D. (1986). *Forces of Production*. New York.

Nuber, Ch. et al. (1989). Werkstattprogrammierung - Setzt sich das Konzept durch? In: *Technische Rundschau* 19. 1989.

Potthast, A. and U. Metzler (1988). Projektierplatz für CNC-Steuerungen. In:

Zeitschrift für wirtschaftliche Fertigung und Automatisierung. 2/1988.

Projektträger Fertigungstechnik (1987). WOP Conference Materials. Stuttgart.

Rosenbrock, H. (1984). Technikgestaltung: Gestaltung ist machbar. In: *IG Metall (ed.). Arbeitsheft zur Humanisierung des Arbeitelebens*. Frankfurt.

Sorge, A. (1985). *Informationstechnik und Arbeit im sozialen Prozeß*. Frankfurt/ New York.

VDI-nachrichten 42.1989.

Vogel, M. R. (1983). *Gesellschaftliche Subjektivitätsformen*. Frankfurt/New York.

Weber, M. (1976). *Wirtschaft und Gesellschaft*. Tübingen.

Weber, C. (1989). Arbeitszeit und Freizeit in Japan. In: *Leviathan*. 2/1989.

Industrial Cultures - Theory and Methods of Cross-National Comparisons

by Lauge Rasmussen

PROLOGUE

'... But fate has decreed that the cloak should become an iron cage. Since asceticism undertook to remodel the world and to work out its ideals in the world, material goods have gained an increasing and finally an inexorable power over the lives of men, as in no previous period of history. Today the spirit of religious asceticism has escaped from this cage - whether finally, who knows? But victorious capitalism, since it rests on mechanical foundation, needs its support no longer'.
Max Weber (1864-1920): 'The Protestant ethic and the spirit of capitalism'.

1 Industrial Culture or Cultures

Cross-national research on the design and production technologies of industrial societies is mostly done on the basis of a comparative functionalistic view of society, treating culture as an independent, external variable which has been brought in to explain observations which seem to be inexplicable from a purely utilitarian point of view.

In reality most comparative studies of technological development fail to develop the concept of culture. Technological change is in some sense autonomous, 'outside society'. Furthermore technical changes cause social changes, which are reduced to 'social effects'. Or it is assumed that social changes determine technical changes in a uniform way that leads to more progress. Technical differences are considered to represent different degrees of progress and progress is more or less thought to advance in the same direction and on the same track everywhere.

The other side of the coin may be just as predictable. Max Weber's metaphor of modern capitalism as an 'iron cage' of bureaucracy and relentless striving after materialistic goods was later mentioned by Jacques Ellul. In his controversial book on 'The Technical Society' (1964), Ellul defines the characteristics of the relationship between the technical phenomenon and society. In Ellul's view Technique defines our epoch in the same way as Catholicism defines the Middle Ages. According to him modern technology is not a factor 'outside society', but a total attitude or comprehensive way of being in the world. Ellul tends to include all human activities within the technological realm. Just like Weber's 'iron cage', Ellul proclaims that

> '... Modern man's state of mind is completely dominated by technical values'[1]

Another well known sociologist Harry Braverman has focused just as singlemindedly on the development of Western societies. In his book 'Labor and Monopoly Capital' he points out:

> '... In this book, we will be concerned entirely with the development of the processes of production, and of labor processes in general, in <u>capitalist</u> society'[2].

Braverman talks of 'society' not 'societies' and makes a historical analysis of the early and late capitalist development without differentiating between the different capitalist societies. Why? Because he is convinced that they have moved and will move in the same direction: towards the state of the 'iron cage' as metaphorically pointed out by Max Weber. In a certain way the 'iron cage' is an observable utilitarian aspect of all industrial developed countries, as Morgan points out:

> '... In a sense, we can thus say that people working in factories and offices in Detroit, Leningrad, Liverpool, Paris, Tokyo and Toronto all belong to the same industrial culture. They are all members of organizational societies. Their work and life experience seem qualitatively different from those of individuals living in more traditional societies dominated by domestic systems of production'[3].

But does 'victorious capitalism' no longer need the support of different cultures? Or are different cultures still alive and in some cases even counteracting each other in modern capitalist countries?

In his book 'Beyond Culture' the antropologist Edward T. Hall supports the last mentioned alternative:

> '... Culture has always been an issue, not only between Europe and Russia, or Japan and USA but among the European states as well. The Germans, the French, the Italians, the Spanish, Portuguese, and English as well as the Scandinavian and Balkan cultures, all have their own identity,

> language, systems of nonverbal communication, material culture, history and 'ways of doing things'[4].

While Weber, Ellul and Braverman focus on the convergent aspects, Hall and the anthropological tradition to which he belongs stress the divergent traditions of culture. One part claims that culture has become uniform or obsolete. The other claims that cultures diverge as they have always done. In my opinion both positions are inadequate. We still need to explore the importance of different industrial cultures even among the technically advanced developed industrial societies. But in order to do this we need a more elaborated concept of culture than the concept that both Weber and Hall reflect. It is not sufficient to reduce culture to a question of 'Protestant Ethic' as Weber does. On the other hand Hall's mixing of 'identity, language, system of nonverbal communication, material culture, history' and 'way of doing things', refers to traditional aspects of culture and not to its transcending or dynamic aspects.

As pointed out by Raymond Williams it is possible to distinguish between at least three levels of culture:

A. 'There is the lived culture of a particular time and place, only fully accessible to those living in that time and place.

B. There is the recorded culture, of every kind, from art to the most everyday facts: the culture of a period.

C. There is also the factor connecting lived culture and period cultures, the culture of selective tradition'[5].

It is the process of selective tradition that makes it possible to understand the dynamic aspect of culture as a continuously selective and interpretative process of learning, communication and production.

Culture is thus not just an absolute body of things and habits of a particular time and space, nor is it only value patterns recorded from a particular period. It is in addition and most important for our purpose - a continuous process of accepting and rejecting areas of what was once a living culture as well as a process of shaping new kinds of relationships between learning, communication and production.

> '... In the analysis of contemporary culture, the existing state of the selective tradition is of vital importance, for it is often true that some change in this tradition - establishing new lines with the past, breaking or redrawing existing lines - is a radical kind of contemporary change'[6].

Weber's image of the 'iron cage' as an unintended spiritless product of the 'Protestant Ethic' is one example of how selective tradition may work towards a particular kind of convergence, if different tendencies are systematically suppressed or rejected.

Hall's interpretation of the divergences of cultures even between European countries can be viewed as recorded as well as lived cultural differences. But we still confront the challenge of understanding the driving forces behind the divergent and convergent tendencies of selective tradition, and the mutual relationships between them.

For example what is the dynamic relationship between divergent and convergent periods of production technology and why do some institutions of industrial societies such as education and communication patterns not seem to converge over long periods.

Why are some industrial countries in a leading position in some technological areas for a certain period and then superseded by other countries? Who are the actors in this play? What role does the institutional background play? Before I go into more detail about these problems, I want to make my general understanding of cultural processes more explicit.

2 Culture as Interpretation of Interaction - the dualistic Tradition

2.1 Weber and the Neo-Kantianism

Weber's studies of the Protestant ethic have been discussed intensively during this century. Could it really be true that a special sect of Christianity has been responsible for the emergence of modern industrial culture in Europe and the United States? According to Weber

> '... The God of Calvinism demanded of his believers not single good works, but a life of good works combined into a unified system ... The moral conduct of the average man was thus deprived of its planless and unsystematic character and subjected to a consistent method for conduct as a whole ... For only by a fundamental change in the whole meaning of life at every moment and in every action could the effects of grace transforming a man from the status natural to the status gratia be proved. The life of the saint was directed solely towards a transcendent end, salvation. But precisely for that reason it was thoroughly rationalized in this world and dominated entirely by the aim of adding to the glory of God on earth'[7].

Emphasizing of the calvinist feature of carrying out the search for salvation in such a way that systematic conduct penetrates all spheres of life to the same extent, the consciousness is prepared for the emergence of modern bureaucracies - and the 'iron cage'. Weber was full of admiration for the organizational accomplishments of modern bureaucracies, but he was sensitive to the price of this development too; when the rationality of bureaucracies was cut loose from the original ethical attitudes and developed into an

> 'inanimate machine ... with power to force men into its service and to dominate their everyday working life as completely as is actually the case in the factory. Objectified intelligence is also the animated machine, the bureaucratic organization, with its specialization in trained skills, its division of jurisdiction, its rules and hierarchical relations of authority [8].

Weber himself was influenced by Neo-Kantianism and its dualistic philosophy of science. He divided sharply between 'is' and 'ought', between statements of fact and judgement of value. This dualism influenced him to consider ideas in themselves as 'cultural value spheres' as being different from 'orders of life' when ideas are connected with interests and institutions. Weber does not draw a precise distinction between the level of cultural traditions and that of institutionalized action systems. But he inspired Parsons and later system oriented social scientists to separate 'culture' as symbolic, interpretative systems different from social actions or interactional systems.

2.2 Parsons, Geertz: Fulfilling the split

Talcott Parsons relied heavily on Weber's account of the transition to the modern age and the development of modern societies. He developed the three-system model of personality, society and culture through which he reconceptualized the legacies of Weber, Durkheim, and Freud in 'functional' terms. According to Parsons:

> '... Culture ... is on the one hand the product of ... on the other hand a determinant of systems of human social interaction. Relative to the particular social system it is a 'pattern' element which is both analytically and empirically abstractive from that particular social system' [9].

In Parson's theoretical scheme, both personality and society are postulated as being interactive systems. Culture is the mediating and stabilizing element of commonly shared systems of symbols.

Deeply influenced by Parsons, the well known anthropologist Clifford Geertz continues the dualistic tradition. According to Geertz:

> ' ... Culture is ... an ordered system of meaning and symbols ... in terms of which human beings interpret their experience and guide their actions, while social structure is the pattern of social interaction itself ... the ongoing process of interactive behaviour ... the actually existing network of social relations [10].

The 'reduction' of culture to being 'pure' interpretation, to being symbols and values 'outside' interaction is fulfilled. Culture is very 'important', a 'determinant' or a 'product' of social interaction, but it is not internal in the process of interaction. By reducing culture to the domain of interpretation abstracted from the domain of interaction itself, it becomes rather tempting to make a further reduction, that is to use culture as a factor 'explaining' what cannot be directly explained by the rational economical behaviour itself.

2.3 Archer: Still in the dualistic 'trap'

In her article 'Theory, Culture and Postindustrial Society' (1990) Margaret S. Archer claims that:

> '... subordination of culture has gathered momentum within industrial society theory, especially over the last decade' [11].

She goes further to argue for this statement by illustrating what she calls three phases in its development, focusing upon their inadequate treatment of culture.

Phase 1: she pronounces the Industrial Convergence Thesis. Following this conviction Clark Kerr (1962) [12] and others believed that industrialism by 'pure logic of the industrialization process' furthered world-wide convergence towards the 'modern society'. It was believed that everyone would sooner or later converge with the United States, and cultural differences would diminish to the extent that a new axial principle of standardized social life would become global.

Phase 2: Post-industrialism. The 1970's versions of Post-industrialism rejected the crude convergence thesis of the 1960's. Instead social structure and culture were reciprocally constitutive of postindustrial society. Though culture in this phase is accepted as an aspect of society it is subordinate to or parasitic upon existing social structures

> '... because of its impotence - it can neither sustain, nor criticize, nor redirect post-industrial society'. [13].

Daniel Bell and Jürgen Habermas are mentioned as the most typical exponents of this phase.

Phase 3: Information Society. The theory of the Information Society was formally initiated by Alvin Toffler [14]. According to Toffler the rapid development of information technology would have deterministic consequences for every aspect of social relation - from domestic life to international competitiveness. Also in this phase Archer found the critics of the Information Society focusing on structural and not cultural aspects:

'... Critical emphasis is exclusively structural and is placed on how we can obtain a better technological mix or fix. Schumacher's 'Intermediate technology' (1978), Roszal's 'Selective industrialization (1973), Illich's 'Convivial production' (1975) and Dickson's 'Alternative technology' (1974). A cultural technique is glazing absent even amongst those who think they are totally hostile. [15].

I found Archer's criticism of the deterministic tendencies in the development theories of industrial societies rather impressive. But she too falls into the traditional 'dualistic trap' when she concludes:

'More is required than an empirical demonstration that the 'Wertrationalitet' is alive and well in modern societies. There is also the vastly important theoretical job which is passed-up by all who endorse the Myth of Cultural Integration of tracing how cultural contradictions within and between belief systems make just as important a contribution to social change as anything going on in the structural domain'. [16].

The question - which Archer does not seem to reflect on - is the relevance of operating at all with separated 'domains' of culture and social structure? Perhaps the question is not to give the cultural 'domain' higher priority than to the structural 'domain', but to understand that the distinction between culture and social structure is false.

3 Culture as Interpretation and Interaction - a Process-oriented Viewpoint

3.1 Where to locate culture?

The dualistic tradition presupposes a particular domain for culture as well as for social structure.

An alternative viewpoint might be that it is quite impossible to locate culture in a particular area. Edward Hall expresses this point very ultimately:

'... there is not one aspect of human life, that is not touched and altered by culture. This means personality, how people express themselves (including shows of emotion, the way they think, how they move, how problems are solved, how their cities are planned and laid out, how transportation systems function and are organized, as well as how economic and government systems are put together and function'[17].

Thus culture is expressive interactions as well as tacit agreements about what is 'relevant' to agree or disagree about. This point is stressed by Sorge too.

> '... there is no culture-free context of organization because, even if organizational solutions or contexts are similar, they are always culturally constructed ... Thus, culture comes into organization through artful, unselfconscious piecemeal experimentation with alternatives in business policy, finance, work organization, industrial relations, education and training, and many other factors' [18].

Culture is here a process of 'experimentation with alternatives' that is a continuous interplay between interpretation and interaction. This is somewhat similar to William's concept of culture as a process of selective tradition, accepting and rejecting existing as well as new ideas and customs. Interaction itself is part of and not 'outside' culture. Culture is not an independent variable to explain all behaviour that is otherwise inexplicable. Nor is culture just a 'product' of art, science or production. Culture is not aimed at a particular area or set of variables, but concerns the interrelationship inside and between all the institutions of a given society. From this point of view analysis of industrial culture should pay greater attention to internal processes as well as interrelationships between the institutions of a society rather than the relation between culture and social structure.

3.2 Institutions as stabilizing aspects of culture

The English word institution derives from Latin, from a verb (instituere) 'to stand something up'. Institution implies that some form of consensus has been 'worked out'. According to Parsons

> '... an institution will be said to be a complex of institutionalized role integrates or status relationships, which is of strategic structural significance in the social system in question [19].

According to Peter Berger, institutions must be viewed from both a micro and a macro social perspective. In the former case, his focus is upon roles as in Parsons' definition. Concerning the macro dimension the focus is upon the structural integration of different institutional sectors. Berger differs from Parsons here. Parsons considers meaningful integration as an objective institutional problem per se. Berger analyses it as a problem of legitimation, justifying in a subjectively plausible way the functioning of social order.

Parsons reflects on culture mainly as a stabilizing part of the social system. According to him, the different national paths of development in different industrial societies indicate the conditions under which structure-forming innovations can best be stabilized. In opposition to Weber's image of the iron cage without the spirit of religious asceticism Parsons proclaims that

> '... In my opinion the Protestant ethic is far from dead. It continues to form our orientations to a very important sector of life today as it did in the past. We do value systematic rational works in 'callings' and we do so out of what is at some level a religious background. In my opinion the instrumental apparatus of modern society could not function without a generous component of this kind of evaluation [20].

Parsons rejects the idea that 'victorious capitalism' (Weber) can rest on the mechanical foundation alone. It still needs men (or women) with the spirit of the 'Calling'. The instrumental, materialistic rationality is not sufficient as a driving force. Weber would not deny that, but left it open:

> '... No one knows who will live in this cage in the future, or whether at the end of this tremendous development entirely new prophecies will arise, or there will be a powerful new birth of old ideas and ideals, or, if neither, mechanized petrification, embellished with a sort of convulsive s-elfimportance. For then it might be said of the 'last men' of this cultural development: '... specialists without spirit, sensualists without hearts'; this nullity imagines that it has attained a level of civilization never before achieved' [21].

Where Parsons proclaims a cultural continuity behind the increasing complexity of social systems, Weber does not exclude the possibility of 'a powerful new birth of old ideas and ideals'.

Since Weber made his 'open prophecy', organization theorists and managers have made powerful attempts to 'regenerate' the spirit of the early capitalism or rather to find a substitute for it. Since Elton Mayo's Hawthorne Studies in the 1920s and 1930s the whole question of work motivation has been a central issue of all influential management theory. Organizational psychologists like Chris Argyris, Frederick Herzberg, and Douglas McGregor began to show how bureaucratic structures could generally be modified to create 'enriched' motivating jobs,encouraging the employees to increase their capacity for self-control and responsibility. Job enrichment, combined with a more participative and employee-centered style of leadership, arose as an alternative to the narrow and authoritarian management orientation generated by Taylor and others. Of course the question is whether these attempts are really alternatives to Weber's 'iron cage' or rather modifications of role definitions which are too restricted even from a functional perspective?

The recent success of 'communitarian' Japan, 'patrimonial' South Korea, and 'patrilineal' Taiwan compared with the decline of important sections of British and American industries have all been crucially linked with the cultural contexts in which they have evolved. For example Murray Sayle [22] (Sayle 1982) believes that the Japanese organizations combine the cultural values of the rice field with the spirit of service of the samurai. The cultural driving force of Japanese industrialization was founded not on a religious ethic but on the preindustrial basic unit of social group formation called ie. According to Murahami:

> '... a basic group unit similar to the ie is rarely found in other agricultural societies. Therefore, I argue that the ie has been one of the main reasons Japan could adapt its indigenous culture to industrialization with extraordinary rapidity' [23].

Sayle as well as Murahami reduces cultural explanations to one or two central institutions of preindustrial Japan. But as Stewart Clegg et. al. correctly reflect:

> '... Behind ... there stood important aspects of an institutional framework, expressed in Japan in terms of patterns of market structure (both labour and capital), union structure and contractual relations which helped to support both elite workers, and elite enterprises with a discernible pattering of corporate culture [24].

In my opinion Clegg et al.'s interpretation of the Japanese economic success so far is correct. Unfortunately, they continue:

> '... however, economic success in Japan ... was less easily adduced to this cultural configuration than to certain technical considerations in production [25].

Clegg et al. continued the reductionistic tradition of culture by interpreting culture as 'groupism' and 'compliance'. They might have chosen another path: Instead of accepting the narrow concept of culture and then rejecting it as a plausible explanation they might have deepened their understanding of the particular relationship between Japanese political, economic, educational and family institutions as a cultural question and thus avoided the dualistic separation between cultural configuration and certain technical considerations in production'.

3.3 The dynamic aspects of culture

Though culture includes important stabilizing aspects as Parsons suggests, dynamic aspects of culture should be considered too. As pointed out by Sorge:

> '... It is important to realize that culture not only has a stabilising function in supplying tacit choices and patterns of action. Through the function of release, it enables people to innovate and transcend received patterns'[26].

While institutions provide a background of role expectations and patterns of behaviour, there is a foreground, where individuals are capable of shaping, innovating and manipulating. Raymond Williams' concept of selective tradition indicates a dynamic aspect of cultural development connecting lived culture and period cultures.

> '... The selective tradition thus creates, at one level, a general human culture, at another level, the historical record of a particular society, at a third level, most difficult to accept and assess, a rejection of considerable areas of what was once a living culture' [27].

The concept of selective tradition is linked to Williams' dialectical perspective of change. While the uniform 'progress' perspective as mentioned earlier restricts our understanding of technology development on the functionalistic level, the dialectical perspective invites us to embrace contradictions as defining features of culture. Thus culture is not only a 'pattern of maintenance' as Parsons suggested.

All phenomena generate latent tendencies and contradictions that tend to change themselves. But opposites are intertwined in a state of tension that also defines a state of harmony and wholeness. This is clearly reflected in the Japanese concept of 'omote' and 'ura', as expressed by Takeo Doi:

> '... Omote and ura belong together although opposing concepts ... Even when we use them separately, one term implies the other: Omote-muki refers to that which is public, open official, ura-muki suggests something private, closed, personal ... Even when we use them separately, one term implies the other: To speak of omote is to speak of ura; to speak of ura is to speak of omote' [28].

According to Doi, institutions - the omote of a society - express the characteristic cultural features of that society. To the extent that the individual - as <u>ura</u> - is expressed by <u>omote</u> institutions, these institutions also belong to the individual. Institutionalization is like a multi-dimensional continuum where the scope as well as the content of role expectations and the integration of different institutional sectors constitute both conflictory and homogenous situations. But this process is dependent on actors. While institutions provide a background of role expectations and patterns of behavior, there is a foreground, where individuals are capable of shaping, innovating and manipulating. The shaping possibility is dependent on 'the function of release' (Sorge). But which elements determine the function of release? This topic will be dealt with in more detail in the following chapter.

4 Perspectives of cross-national Studies

4.1 The problem of change

Our habit of 'thinking in lines' f.ex. technological development as 'progress' - tends to hamper our ability to think dialectically. Consequently we fail to appreciate how the seeds of future technological development are enfolded in the oppositions shaping the present contradictions within a given industrial culture.

As long as anticipations of technical determinism dominated the research of cross-cultural studies of industrial organization, the methods used to select and carry out empirical studies were quite simple. Plants of comparable size and process technology were chosen and studies made, which indicated that technology in the plants of one country was more advanced than it was in the other. But if similar technology was introduced in the 'backward' country, 'deficits' on organizational effectiveness and productivity would be substantially improved. As Kerr et al. (1962) put forward, everyone would sooner or later converge with the United States and cultural differences would diminish to the extent that a new axial principle of standardized social life would become global.

According to Sorge: '... cultural and societal differences were thus used in a residual role in three ways:

1. Where method is concerned, they figured as that part of the variance of international comparisons which could not be explained by well known variables and the interpretation of which was correspondingly problematical.

2. In substance, they were only adduced to explain social-organizational variation to the extent that other known variables proved insufficient.

3. In the projection of social evolution, their importance was played down by the link with different degrees of modernity and by their presentation as being less relevant in the course of further development of society'[29].

This three-fold reduction of culture is a consequence of the dualistic tradition of thinking referred to earlier.

A dialectical viewpoint invites us to recognize that every action has a tendency to produce a movement in the opposite direction. This has major implications for the interpretation of social and technological change. Instead of separating interpretation and interaction or culture and social structure as the dualistic tradition does, the dialectical perspective helps one to focus on how experimentation with alternatives are generated by tensions or contradictions between interpretation and interaction in the existing pattern of social life. But experimentation presupposes actors who are part of certain institutions and are able to transcend the existing state of these institutions in some sense. Precisely the dialectical interplay of 'being part of' and 'transcend' is in my opinion one of the most essential processes for understanding why some industrial cultures seem to be more innovative and flexible than others.

4.2 The dialectical process of innovation - actors and institutions

As stated earlier, institutions imply that some form of consensus has been 'worked out'. Takao Doi has formulated the relationship of institutions and culture in the following way:

> '... just as the face signifies the mind of an individual, institutions, as the face of a society, reveal the characteristic features of that society. If here we consider the concept of culture, and define social institutions as that part of a culture that is enduring and does not change very much it is possible to say that institutions reflect the unique aspects of the various cultures in which they appear'. [30]

I agree with Doi regarding institutions reflecting the unique aspects of the various cultures, however does this mean that they do not change very much? The paradox seems to be, that institutions must change in order to continue to be unique aspects of culture. This may be part of the 'selection tradition' Raymond Williams refers to. Following our dialectical viewpoint every industrial culture contains contradictions and tensions which generate experimentation with alternatives. In certain periods the contradictions are so serious that experimentation is very divergent. New and old kinds of practices are mixed together. Various options and alternatives are considered. Parts of the existing institutions are questioned. But exactly because this happens the more fundamental part of the institution is able to survive. In other periods convergent forces may become dominant. Nevertheless, in the convergent periods the seeds of future divergent experimentation are enfolded in the oppositions shaping the present state of seemingly more and more uniform technology. As pointed out by R-auner/Ruth:

> ' Thus technological development can be interpreted as a cyclical process, in which the divergent elements (influenced by industrial culture) and the convergent elements (influenced by industrial society) are dialectically related to each other and always take effect simultaneously' [31].

In my opinion this process is not a dialectical movement between 'culture' and 'society' as Rauner/Ruth suggest, but a movement inside the industrial culture itself, as well as between different industrial cultures. Cultures contain divergent as well as convergent forces as part of the oppositions created by the dialectical process of certain actors. Institutions provide a rather durable background of role expectations and patterns of behaviour. Actors (groups, individuals) are tied to the institutions either by 'legitimate order' (Weber) or by force. But only to a certain degree. The more contradictions or tensions between the different institutions of a given society, the more likely it is that at least some actors are able to transcend the traditions, and shape new patterns of interpretation and interaction. But what does transcendence means in this sense? Transcendence does not necessarily imply that old institutions are completely broken down and replaced by totally new institutions. This might happen but not very often. Most likely transcendence implies that some innnovations of a social or technical kind change patterns of behaviour without changing the more fundamental parts of the institution. In this kind of dialectical analysis it is crucial to distinguish between primary and subsidiary contradictions. If not, the primary opposition may be concealed by a variety of more superficial differences. Consequently my point is that because cultures comprise many different institutions and different conditions for actors to conform with or transcend part of the institutions and experiment with alternatives, it is absolutely important to make a careful selection of primary oppositions and tensions in order to analyse the process of selective tradition in a proper way. In the following section I intend to exemplify such a process related to an empirical cross-cultural study of production technology and factory organization.

4.3 Primary oppositions of Industrial Cultures of relevance for comparison of industrial design and production

What are the primary oppositions with relevance for understanding national differences of design and use of technology? I suggest three kinds:

A. The opposition between planning and executive tasks
B. The opposition between educational traditions and development of job-structure
C. The opposition between producers and users of production technology

The relationship between planning and executive tasks is often viewed as a question of the degree of hierarchical organization of work, or how far Taylorism is embedded in the dominant organizational orientation of a country. US and Britain are supposed to be more dominated by Taylorism than e.g. Germany and Japan and consequently less able to adopt new flexible strategies on the production level. This is mentioned in Sorge/Warner's study of British and West-German factory organization at the end of the 1970's and the beginning of the 1980's. In this study Sorge/Warner don't talk about opposition, but about 'closeness' and 'overlap' between planning and executive tasks. [32] The same dimension is to be found in Rauner/Ruth's study:

'... The German 'Meister/facharbeitersystem' on the shopfloor level is

> concordant with work/craft oriented hierarchically flat organization patterns' [33].

In both studies it is assumed that flat organization structure is more flexible and adaptable to special kinds of CNC-technology based more on the workshop-oriented concept than Tayloristic oriented organization preferences. I find this hypothesis plausible and worth considering in more detail. From a dialectical viewpoint it is insufficient only to focus on the degree of 'closeness' or 'overlap' as Sorger/Warner do. My point is that a certain degree of opposition between production and design shapes a creative environment for experimenting with alternatives. If the opposition is too overwhelming or if no opposition exists at all, the conditions for innovation of production technology would be more restricted. It might seem to be a paradox that a strong degree of Taylorism may result in either overwhelming oppostion or no opposition at all. The reason behind this 'paradox' should be found in another dimension of industrial cultures: the <u>opposition between educational tradition and development of job structure</u>.

In countries like Germany and Denmark the dual vocational education system is a firmly established institution. Most of the of efforts regarding experimentation with new ways of training is initiated by the vocational education centres rather than by the enterprises themselves. In Japanese factory organization there is a strong institutionalized practice of firm-internal 'multifunctional' skilling by means of cross-training of blue-collar workers (as well as engineers) in different departments as part of their 'carreer-pattern' within the enterprise. In US and Britain 'training on the job' is used but in a much more specialized manner compared with the European continental countries and with Japan. How do these differences in institutional practices influence the kinds and orientations of experiments with new combinations of learning?

Neither management practices nor vocational training and education are independent of the cultural relationships to industrial relations, the work ethics of the youth and the traditional relationships between government ministries, employers' organizations and trade unions. These relationships make it impossible to interpret vocational training as just being an 'appendix' to new technology. There is no direct transaction between the choice of CNC-systems and the kinds of pedagogical networks used to train and update semi-skilled and skilled workers. These networks possess a particular dynamic related to the other institutions. How can we interpret this dynamic relationship? In my opinion Rauner/Ruth's point regarding the different impact of national craft traditions on manufacturing explains a good deal of the different orientations among blue-collar workers as well as white-collar technicians and engineers. But the impact of craft traditions may be a kind of ideology too, legitimizing the 'iron cage' development in the 'disguise' of craftsman attitudes. An example of this is found in Womack et al.: 'The Machine that changed the world' asking workers at a German Carplant what they were doing, they replied:

> '... We're craftsmen, proofs of our company's dedication to quality'[34].

Womack et al. continue:

> 'These 'craftsmen' would have been surprised to learn that they were actually doing the work of Henry Ford's fitters in 1905 - adjusting off-standard parts, fine-tuning parts designed incorrectly so that everything would work properly in the end' [35].

Womack's advice to any company practising 'craftsmanship' of this sort is

> '... Institute lean production as quickly as possible and eliminate the need for all craftsmanship at the source' [36].

Is Weber's prophecy of the 'iron cage' just repeated - but without Weber's scepticism of course? In contrast to Rauner/Ruth, Womack et al. believe that

> '... The great flaw of neo-craftsmanship is, that it will never reach this goal, since it aspires to go in the other direction, (than lean production LBR) back toward an era of handcrafting as an end in itself' [37].

But one thing is car-assembly lines, another is CNC-programming and maintenance one might claim. Is this part of production more adaptable to neocraftsmanship because the CNC operator is likely to have to deal with a greater and more frequently changing range of jobs? This may be an effect of the increased sophistication of the machine control-system through which more flexible change-overs and programmel improvements can be achieved. Another part may be the increasing emphasis on the maintenance functions on the shop floor.

The opposition between educational tradition and job structure is not a question of going 'back toward an era of handcrafting as an end in itself', but a fundamental strength of the innovation potentialities on the production level. Industrial cultures, where this opposition has been accepted as a creative force instead of being eliminated as a disturbing factor, possess the best conditions for shaping and innovating the production technology and factory organization in the future.

The use of blue-collar workers for programming may not only depend on this relationship. Other institutional differences may also be relevant in this case such as availability of skilled or semi-skilled blue-collar workers on the labour market, the industrial relation separation or integration of different occupations as well as the state or private company initiated vocational training for skilled and semiskilled workers. The tendency towards more individualized, customized products, with a greater number of product variants has been associated with the transition from NC to CNC application. Both market-orientated and financial competitive strategies point in the direction of smaller batch sizes and more frequent and less time-consuming machine conversion from one batch to another, arising from the market pressure towards more customized products and components. Furthermore the cost-reduction advantages obtained by reducing finished product stock and work in progress when the full potential for manufacturing 'families' of different, but similar, components is realised point towards the importance of carefully skilled CNC operators. Experience in handling breakdowns, faults, tooling, materials, feeds and speeds is just as important as the programming skills and indicate the importance of keeping the maintenance as well as programming skills on the shopfloor.

Cross-national studies may focus on how these universal trends towards a higher degree of product and component variation are adapted into factory organizations and educational institutions placed in a different cultural contexts? How are the different national kinds of institutional settings and cultural practices able to change or respond to the increasing product-customizing tendencies and to the increasing possibilities of integrating across previously existing departments and hierarchies?

The comparisons should not be restricted to the institutional levels only. As mentioned in my theoretical conceptualization, institutions only provide a background of role expectations and patterns of behaviour. There is a foreground too, where actors (groups, individuals) to a certain extent are capable of shaping and experimenting with new relationships. In some nations or factory organizations the background of interpretations and expectations seems to be much stronger than in others and leave experimentation or creative imagination 'little room'. But exactly this might not be interpreted as a one-dimensional variable of more or less 'room' for shaping and innovations. It may appear to be more complex than that. From a Western cultural viewpoint Japan seems like a paradox also regarding this aspect. The institutional role expectations and patterns of behaviour in Japanese society seem to determine the interpretations and interactions of the Japanese people to a much greater extent than in Western Societies. Thus one may expect a relatively restricted 'foreground' for experimentation related directly to the practical experiences of the involved persons.

On the other hand though, if the institutionally guided patterns of behaviour exercise a strong 'pressure' exactly on using practical experiences to prevent or reduce 'bottleneck situations' or to suggest better use of materials, and equipment or ways of doing things, the 'room' for many small-scale innovations may be more elaborated in Japan than in the Western factory organization.

The relationship between producers and users of production technology is often mentioned as a question of close communication links and customers' degree of influence on the shape of machines delivered. In particular, this is pointed out in Rauner/Ruth's study:

> '... To those societal institutions which are of particular relevance in the German machine tool sector we count the manufacturer/customer relationship. The formal conditions of ordering and delivery, the joint definition of requirement profiles as well as informal manufacturer/customer relations contribute to an industrial culture tradition which allows the customers to have some influence on the shape of the machines delivered [38].

I agree regarding the above mentioned aspects which are also stressed by Binder/Nielsen (36). But I want to add one more aspect. From a dialectical viewpoint the creative forces behind development of new production technology are not only a question of the customers' degree of influence on the shape of the machines delivered. In fact this kind of influence may have restrictive as well as constructive effects on the scope of innovation. Customers/users have no 'patent' of transcending features of production technology. They can as well be tied by their own traditions of organization.

What is important is the way the potential tensions or oppositions between producers and users of production technology are handled. Binder/Nielsen's study shows a very interactive process but not without tensions between producers and users. This study is possibly only applicable for special situations. Nevertheless it is important in order to understand the kinds of complexity involved in the relationship between producers and users. Further empirical studies should focus on this relationship; how oppositions and tensions between producers and users of production technology are either neglected, used in a creative manner or result in new patterns of innovative actors inside as well as outside the plant-level in different industrial cultures.

5 Concluding Remarks

Cross-national studies should not treat culture as an external variable. This means also that the dualistic conceptualization of 'culture' and 'social structure' should be replaced by a dialectical process-oriented concept of culture as presented in this article. It has to be remembered that a dialectical viewpoint is not identical with mutual causality, where 'everything' tends to be the cause and effect of 'everything'. The fundamental core of dialectical development is to focus on how the seeds of the future are enfolded in the oppositions shaping the present. Not all oppositions are primary. In order to prevent a chaotic variety of more superficial dimensions and subdimensions of industrial cultures, one has to select the kinds of oppositions which seem to be most vital to the specific subject, in this case production technology. I have suggested three kinds of primary oppositions related to production technology in this article. Each of these oppositions may be a point of departure for further empirical comparative studies of divergent and convergent forces among different industrial cultures.

In particular, I would suggest that further empirical studies should focus on widening our understanding of the following questions:

A. How does the institutional 'background' of different industrial cultures shape different settings for <u>experimentation</u> with new kinds of vocational education and factory organization? Who are the initiators of the experiments? How are the experiments carried out in practice? How are they legitimized and used? One should not only focus on the stabilizing aspects of culture; it might be just as important to focus on the institutional relationships that enable certain groups or individuals to innovate and transcend received patterns on the shopfloor as well as on design and management levels.

B. Some aspects seem to be more transferable from one national culture to another. In particular some technical systems and organization preferences towards decentralizing responsibilities and job profiles have been transferred but not adapted identically.

The least transferable aspects are supposed to be those which require extensive changes not only in one single department, but also in the related educational and industrial relation institutions. Examples of such changes are management and shop floor practices which are strongly related to traditions of trade unions, vocational traditions or to education-based horizons of problem solving. Is there an underlying tendency towards universal domination of particular factory organization despite the different institutional relationships of the nations - more easily observed in some periods than others - or is the observation of Sorge/Warner regarding a 'core of cultural continuity' more likely?

Is the dynamic between universal and cultural-based divergent development part of a double-loop learning process or better understood like dialectical movements of opposites? Or are other development paths visible too?

Weber's prophecy regarding 'the iron cage' is still open though we know more historical details than he did. Nevertheless it might be fruitful to leave both dualistic and deterministic approaches behind us, and instead focus on industrial cultures as multi-dimensional institutional settings as well as actors of transcending and maintenance of these settings moulded in continuously overlapping cycles or spirals of universal as well as cultural-specific changes. The studies of 'corporate culture' or 'lean production' are far from adequate to catch this dynamic complexity. What is needed, are more comprehensive comparative studies such as those made by Sorge/Warner which are extending the scope to include more explicitly the relationships between the universality of industrial culture and the divergent patterns of adapting those universal trends to particular culturally-based institutions as well as gradually transcending some of the institutional traditions themselves. Studies of the development of design and production strategies related to the particular institutional settings of different industrial nations are well chosen approaches, not only in order to get more detailed empirical knowledge of this area, but also to get theoretical understanding of the relationship between the culturally specific driving forces of institutions and actors in different societies.

References

1.Ellul, Jacques: *The Technological Order* from *The Technological Order*. Ed. Carl F. Stover, Detroit, Wayne State un. Un. Press, 1963.

2. Braverman, Harry: *Labor and Monopoly Capital*, p. 14, Monthly Review Press, 1974, N.Y.

3.Morgan, G. (1986): *Images of Organization*. U.S. Sage Publications, Inc. p. 113.

4.Hall, E.T. (1976): *Beyond Culture*. N.Y., Anchor Press, p. 2.

5.Williams, R. (1961): *The long Revolution*. London, Chatto & Windus Ltd. p. 49

6.Op.cit. p. 53.

7.Weber, M. (1920): *Die protestantische Ethik und der Geist des Kapitalismus*. Danish translation 1972, Copenhagen, Fremad, p. 113.

8. Weber, M. in 'Economy and Society' 3. vol. ed. G. Roth/C. Wittich, N.Y. 1962, p. 1402.

9. Parsons, T. (1951): 'The Social System'. London: Routhledge & Kegan Paul Ltd., p. 15.

10. Geertz, C. (1973): *The interpretation of Cultures*. N.Y.: Basic Books, Inc., Publishers, p. 144.-5.

11. Archer, M. (1990): *Theory, Culture and Post-Industrial Society*, in: Mike Featherstone (ed.): 'Global Culture', Bristol, SAGE Publications Inc., p. 98.

12. Kerr, C. et al. (1962): *Industrialism and Industrial Man*, London, Heinemann.

13. op. cit. p. 102.

14. Toffler, A. (1975): *Future Shock*. London, Pan. (1981): *The Third Wave*. London, Pan.

15. Archer op. cit. p. 109-110.

16. Op. cit. p. 117.

17. Hall op. cit. p. 16-17.

18. Sorge, A., Malcolm Warner (1986): *Comparative Factory Organization*. England: Gover Publishing Company Limited, p. 33.

19. Parsons op. cit. p. 30.

20. Berger, P. et al.: ***The Social Construction of Reality***. Garden City, N.Y., Anchor, 1967.

21. Parsons, T. (1978): *Religion in Postindustrial America* in: *Action theory and the Human Condition*. N.Y. p. 320.

22. Weber (1920): *Die protestantische Ethik* ... (Danish translation. Op. cit. p. 113.

23. Sayle, M. (1982): *The Yellow Peril and the Red Haired Devils* cited from Morgan, G. (1986) op. cit. p. 114.

24. Murahami, Y. (1986): *Technology in transition*. Cited from S.R. Clegg (1990). Op. Cit. p. 41-42.

25. Clegg, S.R., S.G. Redding (Eds.) (1990): *Capitalism in Contrasting Cultures*, Berlin, Walter de Gruyter & Co,, p. 51.

26. Op. cit. p. 51.

27. Sorge, A. (1986). Op. cit. p. 39.

28. Williams, op. cit. note 5, p. 51.

29. Doi, Takeo: *The Anatomy of Self* p. 23, Kodansha International, Tokyo, 1986

30. Sorge op. cit. note 27, p. 39.

31. Doi, op. cit. note 29, p. 50.

32. Rauner, Felix, Ruth, Klaus: *Perspectives of Research in Industrial Culture* p. 119 from *Ergonomics of Hybrid Automated Systems II*. W. Karwowski and M. Rahini (Eds.). Elsevier Science Publishers B.V. 1990.

33. Op. cit. note 30, p. 58 ff.

34. Op. cit. note 32, p. 123.

35. Womach, J.P., D.T. Jones, D. Roos 1990: *The Machine that changed the world*. US: MacMillan Books, p. 91.

36. Op. cit. p. 91.

37. Op. cit. p. 102.

38. Op. cit. p. 102.

The Cultural and Social Shaping of Factory Automation: Towards a New Research Agenda

by J. Martin Corbett

1 Introduction

Social scientists have been investigating the impact of factory automation on work and organisation for many decades and yet no clear picture has emerged from these research endeavours. On the one hand, there is evidence that factory automation (as an independent research variable) has a direct impact on organisation structure (Woodward, 1965), shopfloor skill (Braverman, 1974), production workflow control (Child, 1972), job satisfaction (Corbett et al., 1989), and leadership style (Kipnis, 1986). On the other hand, there is also ample research evidence to suggest that factory automation has no direct impact on any of these variables. Rather, the relationship is crucially mediated by managerial strategic choice (Child, 1972; Buchanan and Boddy, 1986), industrial relations (Hyman and Streeck, 1988), intra-organisational politics and custom-and-practice (Wilkinson, 1985), and other social factors.

Hence, the positions of both technological determinism and technological indeterminism appear to be supported by empirical evidence. Therefore the presenting problem is one of untangling the exact nature of the relationship between factory automation and work (with particular focus on shopfloor job design).

The dominant view of this relationship in the UK is that the thesis of technological determinism is not supported by the evidence. However, this conclusion has encouraged a view of factory automation as inherently neutral and therefore of no direct interest or importance to social scientists engaged in work research. With the increasing adoption of factory automation technologies throughout UK manufacturing and the development of expert systems and automation protocols, it can be argued that this view is both short-sighted and misguided.

The aim of this short paper is to argue the case for the empirical investigation of the nature of the relationship between factory automation and job design by reference to a social constructivist analysis of technology and work (cf. Bijker, Hughes and Pinch, 1987).

The social constructivist approach opens up the social interior of technological design in detail. Even though technology embodies social choices, technology itself has an independent influence on the range of social choices available; i.e. it may constrain social choice (see Clegg and Corbett, 1986; Corbett, 1987). Hence technical design criteria may constitute a critical explanatory variable in an analysis of the relation between job design and factory automation.

A second, interrelated issue concerns the process whereby design choices are made: the conversion of technical possibilities into technological artefacts depends not simply upon technical and market constraints but presumes the construction of collective social alliances (Kidder, 1981).

Thus, the approach of the proposed research agenda challenges the proliferation of case analyses examining the impact of factory automation on work organisation, skills and job design which tend to either read social implications directly from attributes of technological design or read technical design as a direct expression of social interests (Noble, 1984). The social character of design tends to be the outcome of the interplay between diverse forces, rather than the unilateral imposition of power or interest.

2 Research to Date

Research on the impact of factory automation on job design was given a new impetus in 1974 with the publication of Braverman's 'Labor and Monopoly Capital'. Braverman's thesis - that new technology embodies a more effective means of simplifying jobs and enhancing managerial control over the production labour process - heralded a tide of case studies which shows no immediate signs of abating in the UK.

What conclusions that can be drawn from these studies suggest that the context of technology adoption (management strategy, trade union responses, organisational custom and practice, etc.) is a crucial intervening or moderator variable in the relationship between advanced manufacturing technology and job design. Research findings now appear to support a consensus view that technological imperatives are weak, whilst organisational choices are crucial. It has even been stated that

"technology has no impact on people or performance in an organisation independent of the purposes of those who would use it and the responses of those who have to operate it" (Jones, 1982: 199).

As a counter to advocates of unbridled technological determinism, such a statement is perhaps to be welcomed. Yet one has only to look at the well documented reality of working on a car factory assembly lines to see that technology can, and does, have a powerful influence on shopfloor job design (e.g. Beynon, 1974).

With few exceptions the empirical evidence offered as counter-evidence for a technological determinist thesis in factory job design has been amassed through case study analyses. Research in a variety of manufacturing organisations typically shows similar technologies being used in different ways and associated with different operating job designs. In explaining these findings researchers have examined the mediating role of organisation size, culture, politics, environment, management concerns, employee concerns and other non-technical variables. Very few have examined the particular technology under scrutiny, preferring to treat it as a constant or controlled independent variable.

However, the case study approach to factory automation and job design research has five inherent weaknesses. First, it tends to lack a general developmental focus. From the case study literature it is not clear whether research findings on stand-alone CNC (Computer Numerically Controlled) machines can be legitimately generalised to the new breed of FMS (Flexible Manufacturing Systems). The latter constitute a particular configuration of stand-alone CNC machines in which parts and materials transfer is carried out automatically, thereby introducing an additional element of centralised computer control, scheduling and pacing into the job design arena. Similarly, Badham (1989) points out that developments in three dimensional solid modelling on state-of-the-art CAD (Computer Aided Design) technology are tempting the manufacturers of CNC machines towards the design of a new generation of CNC workstations which possess no shopfloor programming facility or capability.

Indeed, the vast majority of case studies have the stand alone CNC machine as their focus. These are the essential building blocks of advanced manufacturing systems such as FMS and, as such, allow an organisation a certain degree of latitude in job design. From case studies of these machines we cannot legitimately build up an overall picture of the impact of advanced manufacturing technology on job design. Littler (1983) shares this criticism of CNC machine tool case studies:

"They are all sampling organisations/firms at a particular historical point, one in which the form of the technology has not been closed off by a series of decisions and technical developments which in combination constitute sunk costs, such that unwinding them - making a series of different choices - becomes an impossible cost burden" (p 144).

A second inherent weakness in the case study approach stems from its essentially atheoretical stance. When, as is usually the case, research findings on CNC technology fail to support a technological determinist line, researchers either replace this interpretation with other forms of determinism (environmental, managerial, etc) or, more commonly, with an open-ended, multi-variable contingency model. Within this research literature, advanced manufacturing technology is often referred to as an 'enabling' technology - one which allows an organisational decision-makers a choice from an increasing range of job design options. This effectively inverts Braverman's thesis that technology tends to constrain organisational choice, but offers no testable theory in its place. Interestingly, the 'flexible specialisation' thesis associated with Piore and Sabel (1984) is typically tarred with the same technological determinist brush which is used to `disprove' Braverman's thesis - despite the fact that the two theses support diametrically opposed hypotheses.

Many case studies cite Braverman's thesis as the key research question and yet they do not directly test the proposition that technology may constrain job design options. This is so despite the implicit recognition amongst many job design theorists that the nature of certain manufacturing technologies (e.g. assembly lines) may overly constrain job design choices. For example, the psychologists Hackman and Oldham, founding fathers of the highly influential Job Characteristics Model of job redesign (1980), argue that

"if work is to be meaningfully redesigned in an organisation either (1) the technology must be of the type that provides at least moderate employee discretion or (2) the technology itself must be changed to be compatible with the characteristics of enriched work" (p. 122).

Similarly Child (1972), a leading exponent of the 'strategic choice' variant of technological indeterminism, admits that

"a given technological configuration...may exhibit short-term rigidities and perhaps indivisibilities, and will to that extent act as a constraint upon the adoption of new workplans (p.6).

A third weakness of case study methodology owes much to the confusing array of definitions of technology (see Corbett, 1987b). It is therefore difficult to compare findings from different case studies. Where measurement has been applied it often fails to disaggregate technological and organisational variables. On the one hand, this is understandable given the widespread development and use of organisation-wide systems technologies which blur the distinction between technology and organisation. On the other hand, given these technological developments, researchers must fine tune their empirical and analytical instruments if the dynamic relationship between technological and organisational variables is to be understood. This is all the more important when researchers are attempting to assess the relative impact of the two variables on job design.

This confusion of definitions is also in evidence within the so-called `new production concepts' debate inspired by the work of Kern and Schumann (1984; 1987). Concepts such as neo-Fordism, post-Fordism, flexible specialisation, systemofacture, human-centred production systems, diversified quality production and computer-assisted cooperative work are used to describe new production concepts and the variable of factory automation becomes subsumed under a welter of ideological, paradigmatic, politico-economic and cultural variables (see Scarbrough and Corbett, 1992).

A fourth problem with job design case studies stems from the treatment of automation technologies as a black box. With the notable exception of studies such as Noble's examination of the development of CNC technology (1984), case studies tend to exclude any examination of the social factors and choices which influence and shape the design of the technology being studied. This failure even to differentiate between different types and makes of CNC machines (and they vary from user-friendly shopfloor programmable machines through to paper tape controlled machines with retro-fitted CNC controllers which have no shopfloor part programming facility) suggests that researchers have established the neutrality of CNC as a generic, homogenous technology before embarking on their job design study.

Thus, the same writers who decry Braverman for his un-dialectical analysis show an equally one-sided bias in their analysis of technology. For example, Buchanan and Boddy (1986) argue that a major determinant of shopfloor job designs around advanced manufacturing technology is the strategic choices made by organisational decision-makers. Whilst this is undoubtedly true, it does not mean that technological determinants are necessarily weak as a result. Advanced manufacturing technology is both a process and a product and there is ample evidence to indicate that developments in technology do not simply derive from technological innovation - they are also driven by strategic choices made by potential purchasers (Roy and Wield, 1986). So, while technological innovations such as the micro-processor may well `enable` the development of shopfloor programmable machine tools and/or CAD workstations fitted with part program generation capability, the ultimate design of the technology will be shaped (at least partially) by perceived market demand and strategic choices made by organisations adopting advanced manufacturing technology.

A final weakness with the case study approach lies in the analysis of job design as a negotiated process carried out primarily between a dominant coalition of organisational decision-makers (i.e. management) and labour. Again, as a counter to unbridled technological determinism, this perspective is commendable. However, this research tends to overlook the possibility that management does not represent a homogeneous group of decision makers. Research by Armstrong (1984) and others indicate that accountants, personnel managers, computer experts, engineers and marketing/sales managers may themselves have conflicting interests and differential influences on the design and ultimate configuration of organisational control systems such as factory automation systems. The concept of 'strategic choice' tends to overlook this dynamic but is one that is readily analyzable through the concept of 'industrial culture'.

In conclusion then, we see that the refutation of technological determinism as a theory to explain developments in factory job design is based on case study analysis. Yet the inherent weaknesses of this approach are such that a determinist thesis remains both tenable and researchable. This is not to decry case analysis as a worthless research methodology. The criticisms above are levelled at the lack of rigour associated with these studies which do not allow a step beyond particularism or analysis rooted in an over-simplified view of Capital-Labour power relations. It is believed that a social constructivist perspective overcomes the inherent weaknesses in the case studies discussed above.

3 Towards a New Research Agenda.

Case study analyses to date have failed to thoroughly examine the cultural and social processes which mediate the relationship between the design of factory automation technologies and the design of factory jobs. What follows is a brief discussion of the key issues which need to be examined to counter this bias. This discussion is based around three rather general premises and is offered as a theoretical catalyst towards a new research agenda for the study of the cultural and social shaping of factory automation and job design.

Premise 1: The technological design of factory automation is shaped by a combination of the technically feasible and the socially and culturally 'desirable'.

Conventional engineering design methods, coupled with designer's perceptions of market/organisational `needs' will fundamentally affect the nature of resultant technological designs. The proposed research will utilise a social-constructivist model of analysis to examine this dynamic (see figure 1).

Technological design does not occur in a vacuum - machines are as much a mirror as a motor of socio-cultural development. Hence, technological design needs be examined as a decision-making process which involves social actors and interest groups who may hold different interests and differential positions of power. Key roles in this process are likely to be those of engineering designers, engineering and production managers, financial and management accountants, computer specialists, and users (and their trade union representatives).

Whilst the political dimension is of key importance here, the role of ideology should not be overlooked - particularly the role of scientific rationality within industrial culture and its application in technological design and use (Rosenbrock, 1990; Scarbrough and Corbett, 1992).

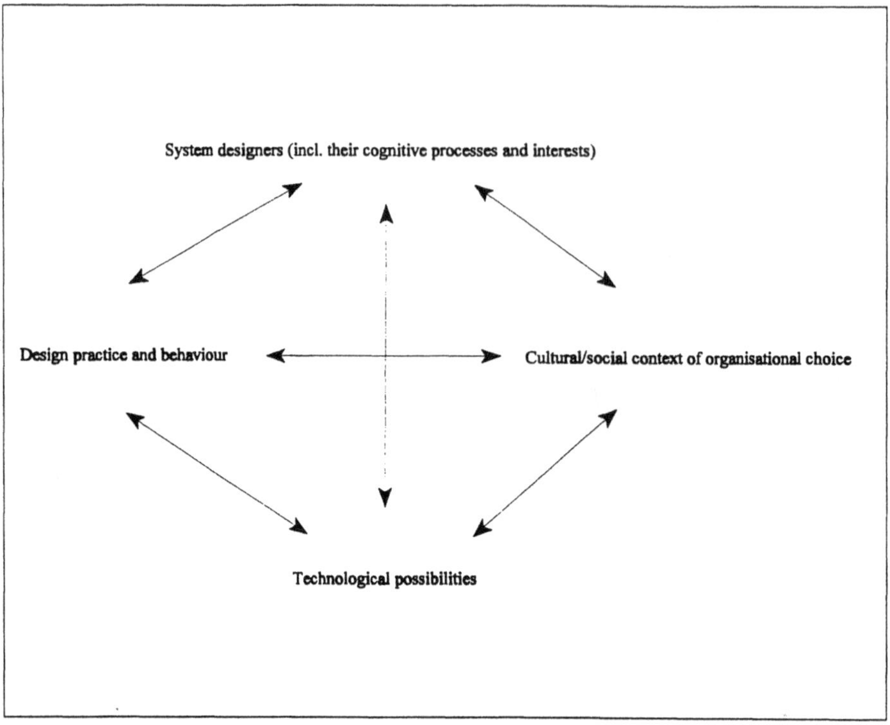

Figure 1: A social-Constructivist Model of Technological Design

In general terms, scientific rationality (as applied to the study of the natural world) is based on direct observation, the uncovering of causal relationships between variables through time and space, experimental methods to test hypotheses, and abstract mathematical reasoning. Using such a base, the goals of science are:

(a) Observation - science is expected to have an empirical base consisting of unambiguous observations of `raw data';

(b) Explanation - science is expected to establish causal links between data variables. Ideally, the variables under study are analyzed in the smallest possible pieces. Atomistic reduction can often reveal how the pieces combine and these combinations are then expressed mathematically;

(c) Prediction - this is the basis of the hypothetico-deductive method. A proposition or theory is tested by deducing that in those conditions covered by the proposition, the predicted event will occur. Hence, an allied goal is that of generalisation from one setting to another; and

(d) Control - i.e. the manipulation of variables. In a controlled experiment specific conditions are created. These are then varied systematically and the range of the experimenters' propositions is tested. Theory is then 'filled in' as the objects of interest are adapted to the progressive logic of the scientists' own calculations (see Hampton-Turner, 1970).

The use to which science and technology can be put usually follows straight from the fact that the its expert practitioners have demonstrated their capacity to predict and control key variables. This feat can be replicated and then disseminated in the form of technical power over identified physical artifacts and elements. Yet, historically, this technical power has extended beyond control of physical objects to the control of people. Whilst, for many, the usefulness of learning to manipulate the physical universe is self-evident, one must ask to whom is the ability to control the behaviour of other people 'useful'? Within a society or organisation comprising social actors and social groups standing in unequal positions of power and influence, those in positions of power are the most obvious beneficiaries.

In this light, Stanley (1978) argues that the encroachment of scientific rationality into technological design and industrial culture has had two profound consequences. First, it has led people to ignore the essential differences and discontinuites between the human social world and the worlds of mathematics and the physical sciences. The ideology of scientific rationality has promoted the view of work organisations as giant mechanisms or machines to which scientific principles and laws could and should be applied. The impact of Frederick Taylor on the thinking of designers and managers of work organisations illustrates both how deeply embedded this thinking had become by the beginning of the twentieth century, and how its consequences became manifest in the world of work.

Miller and O'Leary (1987) argue that by the beginning of the twentieth century, when both engineering management techniques (particularly scientific management) and standard cost accounting were being developed in earnest, the notion of 'calculability' (the use of accounting records as a means for predictive control) became increasingly important within management thinking. Taylorism may be seen as a clear example of this notion couched within engineering terms. For Taylor, the efficiency of an employee cannot be observed with the naked eye and cannot be measured until what is to be regarded as normal or standard has been constructed. Once a standard has been set the employee can become a subject for `scientific' analysis. Similarly, accounting within organisations is about the construction of normalising judgments. "It consists of the task of making visible both within the organisation and in a different form outside the organisation the costs of what people and machines do and comparing this with what they might do. In providing the techniques whereby everything can be evaluated in numerical terms, accountancy provides the archetypal example of the rationalisation process in action" (Morgan, 1990: 100).

A second consequence of extending scientific rationality into industrial culture is that it has led to a belief that any mode of thinking and acting that is not scientific is somehow inferior to thought and action informed by the constitutive assumptions of science. Hence, management will often quote 'efficiency' as a key objective for technological innovation. Efficiency is then defined relative to the quantifiable attributes of technology. In not lending itself to simple measurement, labour (and expert labour) invariably figures as an 'inefficient' variable in management equations. By this logic, labour efficiency increases the more it is predictable, routinised and/or controlled by technology (i.e. more machine-like).

'Efficiency' in the mind of a designer is closely linked to the avoidance of 'lost time'. Indeed, production management often legitimate the exclusion of users in technological design and implementation by arguing that it 'slows down' implementation, and the current fascination with Just-In-Time production techniques equates time with speed of productivity, inventory and work buffer levels, and responsiveness to market demand. Indeed, the control of time (and hence labour) lays at the heart of all current production strategies - whether Taylorist, Fordist or 'flexible specialisation'.

So science and technology are not neutral on two counts. First, when applied to industrial culture, scientific rationality is intrinsically conservative. The need to predict and control produces detachment and an unequal relationship between the observer and the observed, the controller and the controlled. It emphasises the status quo above visions of future possibilities and opportunities. Also, scientific rationality effectively banishes human purpose from its discourse and reduces the person to the status of an object. Ultimately, of course, science is deterministic as our behaviour is seen as being reducible to, and determined by, universal scientific laws and principles. Behaviour that does not conform to these laws (e.g. 'inefficient labour') can then be branded easily as `deviant'.

Second, even the most cursory examination of the history of science and technology reveals that the invention, exchange and use of technology can express and reinforce the interests of the owners and managers of work organisations. This is not to suggest that technological design and use are in some way a direct expression of the purposes of one distinct interest group - the multi-faceted nature of technology makes such control extremely difficult to achieve. Ellul (1974) argues that the encroachment of scientific rationality into social and industrial culture is not the product of some conspiracy on the part of managers, engineers or 'technocrats'. Rather, people are drifting into a subordination to the ideology of scietific rationality, to the conservative values and rationality of science and its application in technological design and use. And, in addition to this general 'drift', scientific and technological rationality remains highly influential in technological and organisational design because it offers techniques which control uncertainty (and therefore grants power to those who can apply in the managerial interest).

Yet it would be wrong to view technological development as in some way entirely beyond the control or power of human agency. This would be an extreme version of technological determinism. Rather, the existence and operation of a scientific ideology surrounding technology means that paths of technological development which contradict or call into question the social relations that reinforce the dominant ideology and culture can only be redirected by actors or groups with great difficulty. Techniques for coping with uncertainty remain an important power resource and organisational groups (e.g. designers, managers, and users) will compete for a monopoly over them.

A case in point here is the demise of "Record-Playback" machine tool control technology. Noble's (1984) study of the development of NC (numerically controlled) machine tools reveals that the rejection of alternative machine tool control technologies (such as "Record-Playback") stemmed, at least in part, from the 'lack of fit' between the latter and the dominant 'predict and control' ideology articulated by higher managerial and executive decision makers within large North American manufacturing and military organisations.

Premise 2: Organisational choice of factory automation design configurations is a function of a social and political process and is not a purely technical process.

This is a more conventional working hypothesis and one which steers much case study research. However, the focus to date has been typically on the socio-political dynamic between management and labour, whereas more research is needed which focuses on the socio-political dynamic between different (often competing) managerial and professional levels and functions, in addition to the management-labour dynamic. There is a need for a focus on the construction of collective social alliances and on inter-professional competition rather than upon any unilateral imposition of power.

Technological innovations evoke opportunities and pressures to change established patterns of power and socio-cultural relations. But, since organisations vary in their cultural and political make-up, the types of opportunities and pressures for change will vary also. Given that the control and use of technology is governed by the social rules embedded in an organisation's industrial culture, the relationships between technology and the pre-existing balance of power in organisations is of key analytical importance.

Whilst technology is subject to the laws of nature, human agents derive social rules concerning its design and use to create what one may term socio-technical systems. Yet, as Burns and Flam (1987) argue, the general tendancy in both capitalist and communist societies has been a separation of most participants from the processes of design, organisation, management and control of these systems.
"Such structuring processes are in the hands of relatively small elite groups. Of couse, designers and producers in many instances, but certainly not consisently, shape and develop such systems according to their perceptions of the needs or demands users and potential users as well as in terms of their own interests and organising principles" (p.301).

Opportunities for powerful groups and coalitions to intervene in the processes of decision-making and implementation are likely to be a function of the scale, complexity and lead-time of the technological change. For example, Wilkinson (1985) describes the opportunities which the debugging of CNC machine tools provides to the shop-floor workforce to influence the way in which the new technology is to be operated. Whilst management exercised their power to purchase and implement CNC technology without consultation with the workforce, the CNC operators were able to exploit their superior knowledge of the technology in use to limit management power to dictate how the technology should be used.

As to the skills and tactics employed in influencing the design and use of technology, Pettigrew's (1973) analysis of a mail-order company's decision to purchase a computer highlights the importance of three key skills, namely;
a) being able to present oneself as an expert;
b) acting as a 'technical gatekeeper' to control the flow of information between top management and outside would-be suppliers; and
c) knowing the system, that is, understanding the culture and balance of power within the organization, and how to get things done.

Pettigrew's study highlights the extent to which an individual manager, exploiting these skills, could effectively dominate the decision-making process on technology despite a wide variety of interests being involved. However, it is important to note that, while the exploitation of such skills may be related to an individual's own ability, much of the underpinning for an individual's success comes from his or her existing organizational position. Pettigrew's study shows how power can derive from a managerial position that allows the individual manager to legitimately claim expertise and control over the decision-making process, and which gave the manager concerned an insight into top management's behaviour.

In this sense, it could be argued that the application of technology is to a large extent a servant of power within organizations; that its application tends to be controlled by (and therefore beneficial to) those in powerful positions. In this view, technology is chosen and applied in ways which continually reinforce the power and control of those in dominant positions. However, this view places too much emphasis on those aspects of technology which are directly controlled or originated by powerful interests (e.g. corporate research and development) and not enough on those socio-cultural elements which lie outside the formal control of a company's owner or board of directors (e.g. the unfolding of scientific and technological knowledge, the use of technology by shop-floor workers, and the operation of product and labour-market forces).

More fundamentally, the notion that technology is simply a servant of power implies that political interests can readily impose themselves on the development and application of technology. But the control of technology is not exerted quite as easily as that - technology is everywhere subject to political processes in which different groups seek to place particular elements of technology under their own control.

In attempting to understand this political process in relation to the development and application of factory automation, the work of Hickson et al. (1971) provides a useful analytical tool. Their 'strategic contingencies' model treats power as a property of social relationships between organisational and occupational groups, which in turn are shaped by factors both inside and outside the organisation. Power is gained and retained through a process of exchange and control of strategic contingencies.

According to Hickson et al., the power of a social group or 'sub unit' within an organisation increases to the extent that it (a) controls or copes with uncertainty, (b) is independent of other groups in achieving its objectives (or conversely that other groups are dependent upon it to achieve their own objectives), (c) cannot be substituted by another, and (d) is central to the functioning of the overall organisation.

In viewing technology as a key contingency in organisational politics, the role of expertise in its application and use becomes paramount. One may view technological development in organisations as a process by which technology is chosen and applied so as to maximise the centrality, non-substitutability and autonomy of a dominant group or coalition of groups. Indeed, it can be argued that technical specialists, as with all professional groups, compete at a group level for access to key positions of command and influence within organisations. In other words, demands for autonomy expressed by professional groups may not simply reflect a desire for autonomy and independence, but an ambition to attain positions of central power and control within organisational hierarchies (and beyond). Layton's (1971) work on the rise of the mechanical engineering profession offers some support for this view.

Layton argues that Taylorism may be seen as an expression of the resentment of American mechanical engineers who, having seen their ownership and/or control of small jobbing shops eroded by the development of big business around the turn of the century, sought a remedy in offering an engineering solution to the key organisational problem of controlling labour. As scientific management (or at least key elements of it) was applied in an increasing number of organisations, the power of the mechanical engineer similarly increased. Armstrong (1985) notes that Taylorism constituted a claim that engineers were the best suited to control big business as it derived its techniques from an engineering analysis of business which engineering experts were better able to apply than any other managerial function.

Although Taylor argued that scientific management could only be installed as a total system, in practice only key elements of it were applied. For example, Taylor's insistence that the control of labour should be monopolised by an engineering planning (or 'thinking') department was generally ignored. As a consequence, engineers were unable to create a monopoly of knowledge and expertise over production. In considering why engineers do not now dominate the management of all major corporations, Armstrong (1985) concludes that the

"techniques of scientific management proved too lucid and could too easily be detached from the ambitions of the engineers" (p.132).

In addition, other professional groups within large organisations (personnel, accountants, and the like) resented the increased power of engineers associated with

Taylorism and Fordism and, in making their own claims for decision-making autonomy, highlighted the shortcomings of an engineering approach to management and attempted to impose their own practices on groups within the production function. Hence, personnel experts could point to the motivational and wider industrial relations problems arising from engineers' Tayloristic notions and offer their own techniques to solve (or at least contain) them.

Similarly, as organisations grew more complex, accountants offered their own techniques for controlling labour through the allocation of budgets and resources. All three expert groups view labour differently as a consequence. To the engineer, human labour is a source of uncertainty within production which must be controlled (or removed through automation); to the personnel specialist labour may be seen as a resource to be fostered; and to the management accountant as a cost to be minimised. This diversity of views has an important bearing on the decision-making processes surrounding the control and use of technology in organisations.

The dynamics and the outcome of this political process is also likely to vary cross-culturally. For example, Armstrong (1985) points out that, unlike the UK industrial culture, industrial management in Germany is dominated by engineers and the techniques of financial control, instead of forming the stock-in-trade of a specialist accounting profession (as in the UK and the US) are taught to these engineering managers at university.

Layton's study of Taylorism highlights two important political facets of expertise. First, expert groups which achieve power within organisations on the basic of solving problems not directly related to the control of labour may still influence management control strategy by virtue of their general dominance in organisational decision-making. A second important political facets of expertise relates to control of knowledge. As part of a strategy for gaining and retaining power, expert groups need to retain a monopoly of their specialist knowledge-base. This may be achieved in a number of ways. For example, the knowledge-base must be sufficiently indeterminate to prevent it becoming part of the knowledge-base of other expert and professional groups.

Additionally, profession groups will take steps to ensure that access to, and the reproduction of, their knowledge-base is kept within their own control as far as possible. This can be achieved by a variety of means. For example, professional associations usually take steps to maintain a certain exclusivity of membership through formal examinations, restrictions on the number of members, codes of ethics and of conduct, etc.

Larson (1977) points out that expert groups retain power not only if they monopolise the knowledge-base and techniques of their trade, but also the evaluation of its end result. For instance, the result of engineering work (e.g. a lathe or an automobile) can be judged by others (the users) whilst the result of accounting work (e.g. a balance sheet) is judged by accountants themselves.

The aim of these tactics of knowledge monopoly is to increase the organization's dependence on expertise by increasing the expert group's autonomy from (and centrality within) the organizational hierarchy, by limiting the extent to which non-experts can define or evaluate the expert group's work.

Thus we see the important political dimension within the relationship between factory automation and work. By going beyond any over-simplified notion of Capital-Labour power relations and into the realm of the politics of expertise, the design, implementation and use of factory automation appears as a complex and fluid process amenable to empirical analysis and investigation.

Premise 3: The design of factory automation technology may constrain choices available to organisational decision-makers vis-a-vis production job designs at the implementation stage.

This hypothesis is one which has not been satisfactorily formulated or addressed by job design researchers and is one which requires the development of more finely tuned measurement tools than have hitheroto been used in case study research. The basis of the research based on this (hypothetical) premise has derived from the `technically embedded constraint' model developed by the author (Corbett, 1987a; 1989).

It is generally accepted that there are political, economic and cultural influences on technological design (see Bijker et al., 1987). Although space does not permit discussion of the ways in which technology design practices are socially shaped, it is necessary to outline briefly the key design assumptions and methods employed by the designers of factory automation systems and to relate these to the concept of technically embedded constraint.

In line with many commentators (e.g. Hampden-Turner, 1970; Noble, 1984; Cooley, 1987), Willcocks and Mason (1987) argue that 'hard' systems thinking predominates in the field of technological design. Hard systems thinking involves the imposition of a clear-cut problem definition on a relatively unstable organisational reality and a `fuzzy' system. It also means the adoption of systematic, orderly, rational procedures that restrict research and design in a very complex organisational reality. A further implication is an overriding concern for technical design and the bypassing of the organisational context in which the system is to operate and the social implications of the system. More fundamentally, it leaves the engineering and computer professionals to decide the extent to which user participation is useful and permissible.

As part of this `hard' approach to systems design, the designer will make (often implicit) assumptions about the way the technical system will be operated by manufacturing personnel. In hard, technology-led systems thinking, where technical considerations are paramount and a high level of automatic computer control is emphasised, the role of the user can only be readily understood and modelled if operating tasks are predictable and well-defined. As a result of these assumptions about human nature and specifications for job design decision-makers within user organisations tend towards a narrow view of human potential and worth. Whilst choice exists at all stages of AMT design, the failure of social scientists and/or users to participate in the making of these choices until the implementation stage can lead to a substantial restriction on the degree of choice over the allocation of personnel responsibilities.

The 'technical embedded constraint' model identifies five key areas of decision-making within the technological design process which have the greatest potential for constraining (or opening up) choice, namely:

1. The allocation of functions between human and machine

2. Systems architecture

3. Control characteristics of the human-machine interface

4. Informational characteristics of the human-machine interface

5. The allocation of responsibilites between operating and support personnel

The model allows the investigation of the ways in which the first four of these decisions may constrain the fifth decision category and is sensitive to different types and applications of factory automation technology. The ideal research sample would contain organisations which use (or are about to use) CNC machine tools, FMS, CADCAM and CAE technologies. This would enable a more developmental approach to be taken in the analysis of this `constraint' process. The model also enables the examination of the extent to which 'hard' thinking predominates design practice and behaviour and the resulting implications for user job design (see Corbett, 1993).

4 Conclusion

The concept of industrial culture has the potential to act as a useful corrective to the over-simplifications inherent in the determinist - strategic choice debate which has dominated UK research on factory automation and job design to date. Cultural and technical variables are mutually interrelated and one of the aims of a new research agenda is to attempt a degree of disentanglement in order to gain more understanding of this complex relationship. To claim that the interplay of technical and cultural (e.g. ideology and the politics of expertise) forces may constrain job design choices is not to assert some kind of 'one-way' determinism. On the contrary, it is apparent that job design (and user reactions to it) plays a key role in the shaping and development of factory automation.

References

Armstrong, P. (1984) `Competition between the organisational professions and the evolution of management control strategies'. In: K. Thompson (ed) Work, Employment and Unemployment: Perspectives on Work and Society. Milton Keynes: Open University Press.

Armstrong, P. (1985) 'Changing management control strategies: The role of competition between accountancy and other organisational professions', Accounting, Organizations and Society, 10, 129-148.

Badham, R. (1989) `Computer aided design, work organisation and the integrated factory`. IEEE Transactions on Engineering Management, 36, 216-226.

Bijker, W.E., Hughes, T.P. and Pinch, T.J. (eds) (1987) The Social Construction of Technological Systems: New Directions in the Sociology and History of Technology. Cambridge MA: MIT Press

Braverman, H. (1974) Labor and Monopoly Capital. New York: Monthly Review Press.

Beynon, H (1974) Working For Fords. Harmondsworth: Penguin.

Buchanan, D. and Boddy, D. (1986) Organisations in the Computer Age. Aldershot: Gower.

Burns, T.R. and Flam, H. (1987) The Shaping of Social Organisation. London: Sage Publications.

Child, J. (1972) `Organisation structure, environment and performance: the role of strategic choice'. Sociology, 6, 1-22.

Clegg, C.W. and Corbett, J.M. (1986) `Psychological and organisational aspects of computer-aided manufacturing . Current Psychological Research and Reviews, 5, 189-204.

Cooley, M.J.E. (1987) Architect or Bee? London: Hogarth Press

Corbett, J.M. (1987a) `Computer aided manufacturing and the design of shopfloor jobs: Towards a new research perspective in occupational psychology' In: M. Frese, E. Ulich & W. Dzida (eds) Psychological Issues of Human-Computer Interaction in the Workplace. Amsterdam: North Holland.

Corbett, J.M. (1987b) 'A psychological study of coupling: the concept of coupling'.

Behaviour and Information Technology, 6, 441-453.

Corbett, J.M. (1989) `Technically embedded constraints on job design and how to overcome them'. Proceedings of IFAC Symposium on Skill Based Automated Production, Viennna, November 1989.

Corbett, J.M. (1993) 'Work at the interface: Advanced manufacturing technology and job design'. In: P. Adler and T. Winograd (eds) Usability: Turning Technologies Into Tools. New York: Oxford Uniuversity Press.

Corbett, J.M., Martin, R., Wall, T.D. and Clegg, C.W. (1989) `Technological coupling as a predictor of intrinsic job satisfaction: A replication study`. Journal of Organisational Behavior, 10, 91-95.

Ellul, J. (1974) The Technological Society. New York: Vintage Books.

Hackman, J.R. and Oldham, G.R. (1980) Work Redesign. New York: Addison-Wesley.

Hampton-Turner, C. (1970) Radical Man. London: Duckworth Press

Hickson, D.J., Hinings, C.R., Lee, C.A., Schneck, R.E. and Pennings, J.M. (1971) 'A strategic contingencies theory of intra-organisational power.' Administrative Science Quarterly, 16, 216-229.

Hyman, R. and Streeck, W. (1988) (eds) New Technology and Industrial Relations. Oxford: Basil Blackwell.

Jones, B. (1982) `Destruction or redistribution of engineering skills? The case of numerical control'. In: S. Woods (ed) The Degradation of Work? London: Hutchinson.

Kern, H. and Schumann, M. (1984) The End of the Division of Labour? Rationalisation in Industrial Production. Munich: Beck Verlag (in German).

Kern, H. and Schumann, M. (1987) 'Limits of the division of labour: new production and employment concepts in West Germany'. Economic and Industrial Democracy, 8, 151-170.

Kidder, T. (1981) The Soul of a New Machine. Harmondsworth: Penguin Books.

Kipnis, D. (1984) `The use of power in organisations and in interpersonal settings'. In: S. Oskamp (ed) Applied Social Psychology Annual 5. London: Sage Publishing.

Larson, M.S. (1977) The Rise of Professionalism: A Sociological Analysis. San Francisco: University of California Press.

Layton, E.T. Jr. (1971) The Revolt of the Engineer. New York: Case Western Reserve University Press.

Littler, C. (1983) 'A history of New Technology'. In: G. Winch (ed) Information Technology in Manufacturing Processes. London: Rossendale.

Miller, P. and O'Leary, T. (1987) 'Accounting and the construction of the governable person'. Accounting, Organisations and Society, 12, 58-71.

Morgan, G. (1990) Organisations in Society. London: MacMillan.

Noble, D. (1984) Forces of Production. New York: Alfred Knopf.

Pettigrew, A.M. (1973) The Politics of Organisational Decision Making. London: Tavistock.

Piore, M.J. and Sabel, C.F. (1984) The Second Industrial Divide: Possibilities for Prosperity. New York: Basic Books.

Rosenbrock, H.H. (1990) Machines With a Purpose. Oxford: Oxford University Press.

Roy, R. and Wield, D. (eds) (1986) Product Design and Technological Innovation. Milton Keynes: Open University Press.

Scarbrough, H. and Corbett, J.M. (1992) Technology and Organization: Power, Meaning and Design. London: Routledge.

Stanley, M. (1978) The Technological Conscience: Survival and Dignity in an Age of Expertise. New York: Free Press.

Wilkinson, B. (1985) The Shopfloor Politics of New Technology. London: Heinemann.

Willcocks, L. and Mason, D. (1987) Computerizing Work: People, Systems Design and Workplace Relations. London: Paradigm Books.

Woodward, J. (1965) Industrial Organisation: Theory and Practice. London: Oxford University Press.

Industrial Culture and Design Methodology

by Thomas Binder and Klaus T. Nielsen

The debate on industrial culture emerges, both academically and practically, from the increasing awareness that design and use of technology reflects beliefs, priorities and orientations in the industrial communities which cannot be reduced to simple means-ends relationships. The evidence stems as much from the appearance of a new plurality in product design on such traditional core-areas of industry as machine tools, as from the fact that close examination of comparable production systems producing similar products, shows significant differences in terms of technical complexity, labour relations and innovative strategies.

One of the questions which can be raised against this background, is how this insight can be adopted fruitfully by designers and users in the process of technological development. In this article we discuss this question in three steps. First, we look into conventional conceptions of machine design and contrast them with the findings of economists and sociologists pointing at significant socio-cultural patterns in the process of design. Second, we turn to the area of systems design, to look into possible parallels for machine design in present trends towards a higher degree of user- and context-awareness in this area. Lastly, we argue the relevance of methodological considerations on how to promote a more conscious and more sensitive dialogue between users and designers concerning both procedural as well as goal related aspects of machine design.

1 Theory and Practice in Machine Design

The concept of industrial culture as it is formulated by Rauner and Ruth (1), is a conceptual framework within which the interpretation of cultural plurality seems both possible and promising. From this framework the points of gravitation of such structures as social institutions, educational institutions, traditions and societal institutions clearly indicate how the innovative potential of the individual design community depends on institutions of regional, national or even in some instances international rooting. The question is however, how design methodology both as part of the engineering education and as a guideline for day-to day design work can make this insight operational.

We find this question is becoming increasingly relevant as conditions and needs for innovation are changing rapidly these years. The general change in market competition away from mass production competing solely on price and towards a more diversified pattern of competition parameters including product quality, product renewal, and customization of products, all indicate a growing significance of genuine innovation at the expense of well known optimization strategies. At the same time, there is a dramatic expansion of technical possibilities in such areas as micro-electronics and new materials as well as a general trend toward a higher degree of externalization of relations between the individual design community and its partners. These trends create an environment for innovation where the stability of institutions in terms of shared views and common understanding can't be taken for granted. What seems to be the case, is that the ability to communicate - the cooperative competence of the design community - is highly decisive for the ability to innovate and compete on the international market.

Focusing particularly on design of production machines, the dual heritage from craft and science has always been evident both for designers as well as for outside observers. Despite this fact the development of technology during most of this century has been conceived as a basically instrumental process, transforming formulated goals into workable solutions. On a theoretical level, economists and other social scientists have taken it as an almost axiomatic assumption that the social process of technological design for all practical purposes could be substituted with the notion that (economic) demand creates the (technical) result. In mainstream design methodology, the same conceptualization has been reflected in a steady trend toward formalizing and rationalizing the design process, giving priority to procedures for identification and selection of choice opportunities. If "traditional" elements have been acknowledged at all in machine design, they have generally been seen as regrettable residues.

In more recent studies of technological development, especially within the tradition of economics of innovation, we have some interesting suggestions for patterns in institutions of innovation which must somehow be retrievable for the micro-process of design. For larger sectors of the economy, Dosi has proposed the concept of technological trajectories as a description for a conglomerate of enterprises pursuing new product and process innovations along the same overall path set out by certain major breakthroughs in basic science (2). In an analogy to Kuhn's concept of paradigms in science, Perez and Freeman have further stressed the socio-cultural origin of these trajectories by suggesting the existence of technological paradigms unifying the development societies around common understandings of "the way ahead" (3). Both Dosi and Freeman & Perez are building on empirical data of major basic innovations and they are, in their line of thought, mainly concerned with the possible relations between cycles in the economy and the building up and exhaustion of what Perez and Freman call "technological regimes".

Pointing to such large movements within the world economy obviously does not in itself indicate a high degree of freedom for search strategies within the individual design process at enterprise level. Nevertheless, the patterns presented in such studies do deviate seriously in their nature from what could be expected with the conventional subordination of technological design processes to economic and managerial ends. The conception of trajectories and paradigms have important explanatory implications for the micro-processes of design in the sense that it indicates a mute concensus in the design communities on roads to be followed, more decisive and restrictive than the mere logic of the market. For methodological considerations the pinpointing of the trend dependency in design in general could open for parallels to areas of design such as fashion or architecture where trend-following as well as trend-setting is an integral part of design methodology.

Historical studies such as those of Sahal give more careful illumination of both diverging and converging tendencies in the process of shaping particular technologies (4). Sahal argues that technological development follows mainly divergent paths in the early stages of the history of a product. In this period, each design community is going through experimentation along a certain set of search lines governed by what he calls technological guideposts. The phase of divergent searching is then according to Sahal, within a rather narrow interval of time taken over by a phase of convergence, where competing design communities seem to relate to the same basic product design, and where design work is focused on optimization. Sahal does not give any in depth explanations of why a certain product seems to gain acceptance as a standard example, but he rejects simple explanations stressing a well defined superiority of the design chosen. In his study of the aircraft industry, he finds the DC-3 aircraft as such an example, which is living on in the consciousness of aircraft designers far beyond the lifetime of the particular technical features of this product. Sahal also finds empirical evidence that different clusters of design communities are able to pursue different lines of innovation in terms of priorities ascribed to different design parameters. In his comparative study of US and Soviet aircraft industry he shows that the US aircraft industry has followed a path to still more light constructions by improvements in strength to weight ratio for construction materials, whereas the Soviet industry has obtained competitive results by improving engine horsepower per unit volume. It is perhaps possible to conclude, also in accordance with Noble (5), that standard examples are likely to appear whenever a critical mass of combined effort is invested in early stages of "history" for that technology. What is most important here is however the mere fact that these examples seem to play a key role for the understanding of the development of technology and furthermore that we in retrospect are able to point to discrete design parameters which both before as well as after the occurrence of a standard example seem to guide innovative search strategies.

From another angle, economists such as Sabel and Lundvall have studied regional and sectoral differences in patterns of innovation. In one comparative study of textile machine manufacturers in Massachusetts and Baden-Wurttemberg, Sabel has argued that rise, fall and renewal in this trade seems to depend heavily on the vitality of user-producer relations, giving an almost reciprocal relation between market position and innovation capacity (6). He studies how US manufacturers were more or less swept from the market by foreign competitors at the emergence of new potentials in the application of micro-electronics. He finds this to be the case despite the fact that US manufacturers had an almost unchallenged oligopoly on the home market, being heavily engaged in expanding their dominance into the textile industry. German manufacturers having the lead in this competition are mainly smaller companies engaged in loose and more egalitarian home market relations with their customers, thus benefiting from customer feedback at the early stages of the development of improved designs.

In a dispute with conventional views of mi-cro-economics, Lundvall has argued that the internalisation of vital customer relations up-stream or down-stream the production chain actually does not take place, but on the contrary that innovations occurring at the interface between cooperating enterprises quantitatively outweigh process innovations (7). Putting user-producer relations in the centre of sectoral analysis, Lundvall has interesting empirical findings from studies of Danish dairy and waste water sectors. Both sectors have well established and stable relations between users and producers. In the dairy sector Lundvall finds an over-average acquisition of new production technology which he ascribes to the monopoly status of the users. He finds that pull for innovation is exceptional for the Danish sector and questions the economic justifications for the sector's investment policy. In the waste water sector Lundvall finds quite the opposite situation. This sector is, according to Lundvall, under-average innovative due to lack of cooperative competence on the side of the using municipalities. Where the dairies seem to pull "too much", the municipalities pull "too little". His conclusion is that not only the organization of user-producer relations in a competitive framework but also the actual competence on both user and producer side seems to decide the quality of innovations.

In contrast to these empirical findings, the present debate on design methodology still has its focus on the decision processes governing the innovative activities of the individual company. Responding to the more turbulent market conditions putting stress on the companies to initiate and implement the production of new products more rapidly than earlier, debates on how to integrate perspectives from production and marketing more closely into the development work has attracted much attention. "Design for ease of production" has become an important key word within this debate. On the operational level it has caused companies to re-construct the development work away from a linear and towards a more iterative process giving production and marketing manegers a more influential position in the associated decision-making. What is striking for the contributors to this debate is however, that they tend to advocate an egalitarian relationship between design and production, but they nevertheless accept the premisses of the managerial structure giving the strive for integration a purely decisionist shape. What could be expected from the patterns unfolding in the descriptive studies of innovation was a more dualistic approach, supporting the complex social processes of learning and searching in the enterprise, along with the development of procedures to manage the fruitful utilization of these processes.

Complementary to the integrative approaches, we see an interest in a methodological approach supporting search processes for product diversification. Contributions in this direction have their origin in psychological traditions of the sixties, where general models of creativity were applied in the field of product design. Like the integrative approaches, these approaches have also had important impact on innovative activities in the companies. The general assumption within this methodological tradition is that the structure of the firm over time becomes hostile to new ideas. Results of the work done within this tradition indicate that their methods are working well for the purpose of distinct renewal campaigns. What is remarkable when reading studies of successful introduction of new products, based on techniques of stimulating creativity among designers and managers, is however that the process of renewal remains external and extraordinary for the companies involved.

Accepting the dualism between the enterprise as production system and the enterprise as bearer of a particular cultural heritage highlights an apparent one-eyed approach of mainstream design methodology. Efforts on methodology have been put into strengthening the converging forces of the system of production, and deals with problems of renewal within a framework of managerial requirements to plan and control. This seems to leave the culturally rooted processes of learning and searching as the absent core to be modified but not acknowledged. In the present situation the vitality of the economy seems at all levels to come to rest on the ability of the companies to utilize and expand on the particularity of their production base. In this situation, it seems likely that new methodological approaches can emerge, which in a more direct manner will focus the attention on social processes of learning and searching both within as well as between cooperating companies.

2 Design of Computer Applications

When discussing future trends in design methodology in machine design, developments in the field of designing computer applications are interesting in several ways. First of all, computer science emerges from the same tradition of modern engineering, which also forms the basis for the self understanding of developers within design of production technology. The systems concept together with the fundamentally mechanistic approach to the task of designing were constitutive for the formation of first generations of computer technology and opposite to machine design, the traditional elements inherited from the crafts tradition were absent in the design of computer applications. As could be expected, computer science for long has delivered some of the strongest concepts to be applied also in other areas of design. But where this is still true when looking at such concepts as computer aided design, computer numerical control and at a higher level the entire notion of computer integrated manufacturing, it is now interesting to see how designers of computer applications have turned their attention increasingly towards the users and their environment.

This interest in users and user context is, in our view, the result of increasing difficulties with formalized methods of design, such as the methods proposed by DeMarco and Yourdon (8). These difficulties include lack of the ability to adapt computer systems to the tacit skills of the users as well as the general mismatch between design of closed systems and the organizational needs for adaption and adaptability. Stimulated by the development of new programming tools, the focus in the development process has generally changed away from technical issues concerning programming and towards user aspects of the systems. This implies together with the trend of user involvement, a shift in orientation among designers. In general, the perspective is changing from stand alone systems involving individual users to integrated tool-like systems supplying a collective user-group with new abilities. Corresponding to this trend, we see changes in the design process. Prototypes and prototyping as design method gains ground and the continual modification of prototypes ending up with the final product is an important shift away from a linear and towards a more iterative and responsive approach to design.

It is still too early to talk about a new tradition in mainstream design of computer applications. The discussion is however vivid and a number of groups have elaborated on the new trends in order to establish a more comprehensive and reflected design methodology incorporating user participation and use-orientation as central categories of design (9).

The question of designing user-interfaces is one of the questions which seems to be changing status at present. In a classical systems design approach, the interface question is basically a question of mediation. The system as such has a functional core, which the user has to handle adequately. The control of nuclear power plants has been an important field for such user-interface considerations and in our view has had some of the qualities of what Sahal calls a standard example. The problem of designing user-interfaces for such systems has been how to display, for the user, the functionality of a control system which is essentially automatic. Several studies in the field have shown that this is a delicate question. Users in general experience the nature of a particular system through its use and have difficulties in obtaining action-taking competence through formalized representations of a functionality which remains opaque to them. Focusing on other computer applications with less overwhelming functional dedication such as systems for processing material for publishing, a number of authors have questioned the necessity of distinguishing between system core and user-interface. Bødker has argued that the designing process should take the user-interface as its starting point, and as she formulates it, design the total application "through the interface" (10). Her point is that for a great number of applications such as systems for newspaper paste up, the functionality of the application is intimately connected to its use/user properties, and therefore has to grow out from "the fingertips of the users".

Bannon and Kuutti has a somewhat similar argument in a discussion of how designers can deal with the fact that designing of computer applications seems so intrinsically interwoven with the construction of new work practices and patterns of communication at the user site (11). Their argument is that the task put forward to the designer cannot be adequately solved without subordinating the work on the application to the considerations about the future workplace. Giving this argument in the context of the debate on human computer interaction, Bannon and Kuutti aim at broadening the scope of traditional interface considerations. On the one hand their argument is that functionality and user-interface of an application cannot be adequately separated because the task of the design work is to produce use functionality as it can be experienced by the user. On the other hand they do in a more general sense question the drawing of the "interface line" between application and user context by stressing the fact that the resulting usability stems as much from construction of context as from the particular design of the application.

Another question concerning the targeting of methodological considerations arises from the fact that designers increasingly employ some form of prototyping and prototype demonstration as part of a dialogue with the using organization. Grønbæk has studied prototyping activities in a selected sample of leading Danish application design enterprises (12). He finds that these enterprises use prototype demonstrations to get reactions from future users, and often put considerable effort into these demonstrations. He also finds that prototyping as design method, with successive simulations of the finished product is widely used. Whereas these observations indicate a shift away from the linear design process propagated in the early years of computer science, Grønbæk finds important unsolved problems in the way the companies engage in the dialogue with the user organization. From observations during prototype demonstrations (13) he finds that in general reactions from the users are few and of a certain random character. This, he argues has to do with the properties of the prototypes as well as with the set up of the dialogue. Most of the companies studied use, what he calls horizontal prototypes with no or only very limited functionality. This means that the future users are presented with what is essentially a screen design, and have to imagine from there how the application will be to use. He argues that prototypes have to be vertical with a considerable degree of functionality to allow the user to go through a simulation of well known work processes, and hereby experiencing the pros and cons of the design. Users most go through simulations of the future work process and find the limits and possibilities of the application to be able to fully contribute to the design process. He stresses particularly the situations of "breakdowns" where the user expects something to be possible, but is disappointed, to be especially fruitful. In these situations the user experiences what she expects, and is brought to consider how the difference between expectations and realized application is to be dealt with.

What is important for Grønbæk and for us as well, is the indication that a user-designer dialogue has to build on the opportunities of experiencing possibilities in practice-like situations where both past work practice as well as future design options are brought into the dialogue. In explaining why the companies do not seem to obtain more useful contributions from the users in the user-designer dialogues studied, Grønbæk points to two aspects of the design practice. First, he points to the ambiguous position of the designers, wanting on the one hand to gain feed back from the users, but hoping on the other hand to get acceptance of the design put forward. The need felt to involve the users still conflicts with the conventional aspiration to get the work done. Second, he and others have the interesting observation, that the possibilities of designing and re-designing user-prototypes with moderate investment of resources tend to narrow the considerations of alternatives to the range of possible solutions inherent in the prototypes modelled early in the design process.

What we find especially noteworthy in this on-going debate is that computer scientists as such seem to have changed their course from being the agents of formalization and systemization per se, to be the more humble and context-searching supporters of the existing organization. Although the orientation towards use context and users is still ambiguous, we see in this fact also a new trend in the fraternalistic dispute between software designers and designers within mechanical design. The concept of prototypes and prototyping has actually borrowed a lot from traditional methodology within machine design - and though systems design still has the upper hand in shaping the ideal of design in modern engineering, the day-to-day search for context among application designers actually plays the ball back to design areas, where tradition is a constitutive part of the cultural heritage and as such is still a part of practical design activities.

3 Towards a Language Orientation in Design Methodology

While preparing this article, we had discussions with a number of people presently engaged in the Danish debate on design methodology. One of the more striking comments we got when we presented our idea that machine design could learn from the debate among computer scientists on user involvement and context awareness, was a criticism saying that our point of view overlooked the fact that designers in the field of machine design, unlike the computer scientists, have always been rooted in a long tradition of how to do things. What computer scientists are now searching, one of our opponents argued, is basically a compensation for the common "know how" in mechanical engineering.

This remark is particularly striking because it comes from one of the leading propagators of what we earlier in this article called "design for ease of production". To a large extent, we accept the argument, but are uneasy to accept it as a critique of our own point of view. What our opponent is actually stating, is that he is also aware of the socialization to tradition as a formative ingredient in the shaping of any particular design community.

Our point of view is that "design for ease of production" like most other present contributions to the debate on methods in machine design underestimates the traditional elements of the design practice in the search for a "scientific" methodology. As stated earlier, this approach becomes particularistic in its focus on the managerial needs and the apparent neglect of culturally rooted learning processes as a core in all design-activity.

Furthermore, our point of view is that to accept the binding to tradition as fundamental for design is not contrary to involvement of users in the design process. Also in machine design where tradition is vivid, acknowledged or not, an expansion of the dialogue between user and designer will be an improvement of the design process because it makes the elements of tradition conscious to the designers. The elements of tradition now being from both the side of the designers and the side of the users.

In our own preliminary investigation into the Danish sector producing customized special purpose machine tools, we found that problems concerning the dialogue between producer- and user-companies played a major role for designers understanding of their work. The producers being mainly smaller companies innovating around a certain basic machine design agree on a certain development task on the basis of requirement specifications set up by producer and customer in common. What they experience then is that the customer company does not realize until late in the process, what they actually see as their production need. In general, the companies seem to engage in development contracts with the mutual expectation, that the designers know the problems of the customer company - when this expectation during the development work is questioned, the lack of experience in communicating needs, possibilities and constraints tends to provoke legal conflicts rather than common dialogue.

"Scientific" methods of design might first have proved to be superficial and inadequate in the area of design of computer systems because of the lack of an underlying pragmatic tradition. This may explain why designers of computer systems have moved ahead in the search for operational ways to conduct the user-dialogue. In our opinion, design methodology in machine design however has lessons to learn from the methodology in design of computer systems, because conditions for machine design are getting turbulent, and the design goals are becoming just as open as we know them for computer applications.

In the past few years, the notions: "users" and "user involvement" have changed status from being key words in criticism of mainstream development work to becoming an all-embracing and highly diluted headline for almost all kinds of demonstration activities. This change stems no doubt to a large extent from the fact that design communities in general have become more aware of the need for adaptability to and acceptance in the implementation environment. However, what is so nicely demonstrated in the empirical study of Grønbæk, that contributions from the user side presupposes a considerable degree of exploration and dialogue, in our view clearly highlights the limits of such generalized insights. One of the ice breakers for the understanding of why users have to be involved in the process of technological change has been the debate around Polanyi's concept of tacit knowledge. A number of theoretical and empirical studies on the resulting work practice after installation of highly automated equipment have shown that rather than deskilling the outcome seems to be a continued or even increased dependence on the skills of the workforce. The skill deemed inaccessible for formalization and subsequent automation has been termed tacit, and this residual skill category has gained an almost mythical status in parts of the working life debate. Despite the valuable contributions delivered along this line, we find that the concept itself to some extent stands in the way of establishing fruitful relations between designers and users. Our point is that users shall be seen as end-users and that user involvement shall mean the participation in mutual learning processes between users and designers. Furthermore, the engagement on the designer side in user-designer cooperation is not as much a question of lending an ear to the voice of the user, as it is the installation of a forum where views, experience and visions can be presented and reflected upon by users as well as by designers.

In systems design, the question of what to discuss with the users is closely associated with the question of where and how to draw the "interface line". Another way to formulate this question is to discuss whether to design from a tools perspective or a systems perspective. That is whether the end-user sees the system as a tool or rather as a system where he or she just plays a minor part. In machine design this question corresponds to a similar tension between accepting and supporting the existing competence in the user organisation, or pursuing systemic rationalization neglecting existing competences and practices. Although this tension is somewhat beneath the surface in mainstream machine design, we see an emerging trend towards reconsidering the conceptual foundation for machine design from the pioneering work of Rosenbrock to the present debate on anthropocentric production system propagated by such authors as Brödner, Corbett et al and others (14).

We think the question of where to place the user dialogue in relation to the entire design process can be answered much along the line of the considerations in computer science presented earlier. In machine design man-machine considerations is well known from industrial ergonomics, but also here the drawing of the interface line seems to rely on a basic consensus on the functionality of the machine which cannot be assumed when broader use characteristics are questioned. Where the interface considerations and particularly the design through the interface approach, in computer science, to some extent serves as argument for a prototyping methodology, this have to be further detailed when discussing machine design. As we have already mentioned, the use of prototypes in the design work is a well established practice in machine design. What seems relevant in machine design is to explore, more fundamentally, the possibility of reversing the sequence of design considerations in order to start from the use functionality and gradually move inwards to the inner functional properties. Use model prototypes which can be discussed upon from a user perspective, and just as important some kind of opening of the designers toolbox for the user seems to us to be important methodological goals.

Finally the dialogue in itself needs support from methodological considerations. As the user-designer dialogue is basically aimed at supporting reflection both on the user and the designer side, we need formats and concepts which on the one hand makes it possible for users and designers to communicate experience, needs and possibilities, and which on the other hand avoids the traps of objectifying formalization. In dealing with this aspect, we see the possibility of a more generally applicable conceptualization of the entire design process. Inspired by Ehn's notion of user-designer cooperation as a meeting point between distinct but overlapping language games (15), we think that the understanding of design as basically communication, delivers a useful perspective for design considerations. Methodological recommendations can from this perspective take a linguistic turn. Ehn and Sjögren have proposed to see the design process within a theatrical metaphor (16), where users and designers are playing their way through possible work situations. Both users and designers are actors experiencing potential benefits and constraints without spelling out arguments of opposing views. Corbett, Rauner and Rasmussen have suggested a workshop model for joint design work for designers, users and social scientist, employing scenarios as the format for presenting and evaluating concepts of future designs (17). They propose a mixture of homogeneous and mixed work shops for the participants in order to allow both the creation of new suggestions and the critical consideration of suggestions rooted in past experience. Along a similar path, Kensing & Madsen has suggested dialogue around different metaphors applied on the context for the design work to support reflection that transcends the views and experience embedded in the established practice (18).

In our preliminary work on design methodology for the design of customized special purpose machine tools, we have proposed the development of working documents to work on and to keep track of the process in user-designer cooperation. We suggest three basic documents: the designer handbook, the user handbook and the user-designer protocol. The designers handbook is consisting of building blocks for the design, derived from previous design processes, and modelled in a way that makes "use-model" modelling accessible for the users. The users handbook is consisting of problem and context scenarios, modelled in the perspective of physical structure, organizational structure and time structure of the work place. The user and designer handbooks are tools for users and designers to pin-point areas of problems or possibilities when evaluating use-models of future artifacts (as well from a skill perspective (the artifact as a tool) as from an organizational perspective (the artifact as part of the working environment)). The user-designer protocol is then a procedural guideline, establishing a format for the documentation of considerations and decisions taken when design is in progress. The handbook and protocol concepts are inspired by the theatre-metaphor proposed by Ehn & Sjögren. The handbooks are documents developed by the designer and user respectively and they can be seen as reminders for the role play in the common design process. Correspondingly, the protocol forms a basic script for the dialogue, highlighting themes to be explored and conclusions to be arrived at.

The suggestions presented here only indicate a direction for methodological consideration. We find however that the orientation towards language and the attempts to install the user designer dialogue firmly in the core of the design process may make the way open for the design communities to fertilize the potential in the inevitable user-producer relations.

References

1. Rauner & Ruth: "Perspectives of Research in 'Industrial Culture'" in Karwowski & Rahimi (eds.): "Ergonomics of Hybrid Automated Systems II", Elsevir, 1990.

2. Dosi, Giovanni: "Technological paradigms and technological trajectories", Research Policy 11 (1982) pp. 147-162, North Holland publ.

3. Freeman & Perez: "Structural Crisis of adjustment: business cycles and investment behaviour" in Dosi et al (eds.): "Technical Change and Economic Theory", Pinter Publishers, London, 1988.

4. Sahal, Devendra: "Patterns of Technological Innovation", Addison Wesley, 1981.

5. Noble, David: "Forces of production", Knopff, 1987.

6. Sabel, Charles et al: "Regional Prosperities Compared: Massachusetts and Baden-Württemberg in the 1980's", discussion paper, Wissenshaftszentrum Berlin, 1987.

7. Lundvall, Bengt Åke: "Innovation as an interactive process: from user-producer interaction to the national system of innovation" in Dosi et al (eds.), 1988, op.cit.

8. DeMarco, Tom: "Structured Analysis and Systems specifications", Prentice Hall, 1979, and
Yourdon, E: "Managing the Systems Life Cycle", Yourdon Press, 1982.

9. Greenbaum & Kyng (eds.): "Design at Work: Cooperative Design of Computer Systems", pre-published version, Aarhus University, April 1990.

10. Bødker, S: "Through the interface - a human activity approach to interface design", Lawrence Erlbaum Ass., 1990.

11. Bannon and Kuutti: "Some confusions at the interface: Reconceptualizing the 'interface' problem", paper at the conference "Human Jobs and Computer Interfaces", Tampere, Finland, 1991.

12. Grønbæk, Kaj: "Rapid prototyping with fourth generation systems - an empirical study", DAIMI, Århus, 1988.

13. Grønbæk and Bødker: "Cooperative Prototyping experiments", DAIMI, Århus, 1989.

14. Rosenbrock, H.H.: "The future of Control", Automatica, Vol. 13, 1977.
Brödner, P.: "Fabrik 2000", Wissenschaftszentrum Berlin, 1986.
Corbett, Rauner & Rasmussen: "Crossing the Border", Springer, 1990.

15. Ehn, Pelle: "Work-Oriented Design of Computer Artifacts", Arbetslivscentrum, Stockholm, 1988.

16. Ehn & Sjögren: "From Systems Descriptions to Scripts for Action" in Greenbaum & Kyng (eds.), 1990, op.cit.

17. Corbett, Rauner & Rasmussen, 1990, op.cit.

18. Kensing & Madsen: "Generating Visions: Future Workshops and Metaphorical Design" in Greenbaum & Kyng (eds.), 1990, op.cit.

Industrial Culture and Software Production

by Jürgen Friedrich

1 Software Technology, Culture, and Production

1.1 Information technology: Forces of development

The development of information technology is normally explained in terms of various "economic forces": the strength of competition, the global market place, the aim of maximising profit etc. Of course economic rationality is one of the main factors responsible for the break through of technological innovations. But this one-dimensional approach cannot adequately explain the complex framework of technological change, especially in the information technology sector. Aspects of non-economic forces of development have also to be taken into account: Technological, political, societal and industrial-cultural factors are influencing the technological innovation process as well and are not to be underestimated in the theoretical analysis of technological change.

1. There seems to exist an inherent *technological logic* of the innovation process. The invention of the 16 megabit chip has been followed by the search for a 64 megabit chip, without any questioning of the wider issues by electronic engineers. Miniaturisation, signal transmission rate, and memory size are going to become values per se. (It would be the task of a separate paper to give an elaborate analysis of the highly ideological character of the concept of "technological innovation logic".)

2. Most countries as well as several supra-national institutions are enforcing the technological innovation and implementation process by political means such as tax allowances on research investment, direct research funding or infrastructural measures in the scientific as well as the vocational training area. By this *political logic* the aims of the single enterprise are transformed into strategies at a national and increasingly global economic level. (There is at least one more political factor heavily influencing the development of information technology: The military demand for complex communication and control systems. A deeper analysis would show mutual interrelations but also contradictions between the economic and military logic of development in the field of information technology.)

3. *Societal factors* as determinants of technological development are often neglected as well. Habits, traditions, orientations, and attitudes of people towards information technology differ from country to country and influence the "innovation climate". Societal factors are to be found at different levels of abstraction. They range from the question of consensus about the nature of technology, e.g. the belief that technological and social progress are inextricably linked, to the personal identification with technical products, e.g. the degree of penetration of consumer electronic products in private

households. In the present context there is no great difference between societal and *cultural factors* of technological development: Culture (in a static sense) is seen as a shared understanding of a group (society, nation) concerning basic knowledge, values and rules that influence our interpretation "of the world" and the behaviour of individuals and groups "within the world". Culture (in a dynamic sense) also covers the process of evolving a shared understanding by ongoing communicative acts, a continual discourse based on conflict and consensus.

4. Transferring the concept of culture to working life in modern industry, *industrial cultural factors* are constituted by the economic, organisational and social relations of the participants in the production process. In the article paper we focus on these industrial cultural determinants which have been widely ignored as influencing factors in the history of information technology and especially of software development.

1.2 Industrial culture

At first sight "industry" and "culture" seem to form a contradictory relation. Industry describes the domain of technology and production, work and economy, whereas culture is associated with arts and sciences, recreation and lifestyle. But as we mentioned above, culture in sociology has a much broader sense, so the precise meaning of industrial culture can be understood as the social framework of mutual behaviour in work and production. We wish to follow the definition of industrial culture given by Rauner and Ruth (1988, p. 7):

"By 'industrial culture' we understand all those factors that determine the specific shape of any actual technology in a dialectic interplay with the processes of technological development. 'Industrial culture' comprises a country's specific industrial relations as well as the corresponding strategies and traditions in carrying out social conflicts, the specific institutional organisation and qualification of the labour force as well as national structures of the division and organisation of labour."

In particular industrial culture is formed by various attitudes, interpretations, strategies and organisational schemes of which we would like to mention the following (cf. Rauner/Ruth 1989; Hildebrandt/Seltz 1989):

- Management techniques
  (autocratic vs. participatory style)

- Preferences concerning work organisation
  (degree of the division of labour, degree of hierarchical structures)

- 'Social contract' at company level
  (system of informal rights and obligations concerning policies of employment, continual training, wage fixing etc.)

- Industrial relations
(co-operative/conflict-oriented relations, bargaining processes, collective agreements etc. at company and branch level, work legislation on national level)

- Educational system and vocational training system
(qualification schemes, apprenticeship, degree of semi-skilled labour, state intervention)

- Professional orientation of the employees
(attitude towards work contents, social assessment of work and leisure etc.)

- Problem-solving perspective and design orientation of systems analysts and engineers
(human-centred vs. technology-centred perspective)

So far results concerning the impact of industrial cultural determinants on the development of technology were mainly gained from research in the field of material production, especially with respect to flexible automation. For an adequate understanding of the processes in this specific application area but also in general, it is also necessary to investigate the underlying *basic technology*. This basic technology is information technology, consisting of microelectronics and communication technology on the one hand and software systems and edp-organisation on the other. Software design and edp-organisation is - this is my first hypothesis - more heavily influenced by industrial culture than is hardware. What are, in the field of software and organisational design, the key concepts of cultural divergence between the leading approaches in present computer science? Before we try to answer this question we have to clarify the double interaction between "software" and "production".

1.3 Double-minded: Production software and software production

In the context of software development and application, "production" means on the one hand the *area* where software is applied to control a variety of machine tools and provide the infrastructure of data communication at shop floor level. On the other hand the term "production" denotes a special *methodology* of software development which is derived from the principles of the material production process.

Let us consider the first aspect: Why is the question of cultural determination of software technology so important with regard to automation of production? It might be stated - this is my second hypothesis - that in many cases the possibility of a social shaping of production technology is significantly constrained by the previous design of the underlying computer technology, and especially by the applied software methodology. I shall try to illustrate this hypothesis by offering two examples: a) Computer science - descending among others from "mathematical culture" tends to formulate all operations and conditions in terms of formal calculus. Therefore programming languages are - without any inherent necessity - formal (symbol manipulating) languages. What are the consequences of this for the programming of machine tools on the shop floor? From the viewpoint of computer science e.g. teach-in-"programming" is only a short period of "passing through the dark tunnel of ignorance to the light of numerical control programming".[1] b) The batch area of computing resulted in a division of labour between programming and use of edp. This concept was implicitly transferred to the production area, providing the basis for the dominance of a specific organisational form in applying the CNC technology, namely the division between technical office (programming) and shop floor (operating). As we know this organisational form is not at all a necessary precondition for the application of CNC machine tools, but is probably the result of a programming tradition which is heavily influenced by modern computer science thinking.

In so far as information technology is a substantial precondition for the automation of production today, the basic assumptions about work which are implemented in the problem solving methodology of computer science are transferred to the application area. So, the key for a social shaping of automation technology up to a certain extent lies in the hands of software engineers and depends on the process of model building in computer science.

The second aspect of the interrelation between software and production refers to the methods software is generated by. Which term best describes this generation process: Design, development or production? Each of these terms is associated with a specific analogy to the field of material processing. Design and development focus on the creation of a single or even unique item. Production however bears the meaning of a repeated application of identical steps out of a limited repertoire of process elements. Regardless of whether this explanation of the term production is correct from the viewpoint of the methodology of production technology, it reflects - as a metaphorical expression - a special orientation of software development which is connected to the factory concept (cf. chapter 3).

So, in the software development debate "industrial culture" has been interpreted in two ways: *Industry* stands for standardisation and mass production, *culture* is associated with creative design and "one off programming". Actually both software development philosophies are found in practice. The question is, what are the consequences resulting from the respective approach in terms of social and economic impact, in other words: with respect to humancentredness and competitiveness.

---

1 The history of machine tool programming is at present being repeated in the development of robot programming languages.

2 Industrial Culture I: The Culture of Software Design

2.1 The mental orientation of software developers

It is an illusion to suppose that the development decisions of software designers are controlled by clear organisational guidelines, such as formal specification methods, programming environments, documentation rules, etc. As early as 1974 Hedberg and Mumford stated: "It is reasonable to assume that values play an important part in guiding the designers' choice between different design alternatives." (Hedberg/Mumford 1974). Values are constituted by the individual background and the organisational context of the designer. The individual background is formed by the designers education and vocational training and thus strongly depends on the code of conduct within the professional community. Furthermore the software developers seem to identify themselves with their ("modern", "promising") work to a greater extent than other professionals do. This fact emphasises the influence that the developer's personal design orientation has compared with organisational rules. The organisational context consists on the one hand of the value system of the enterprise, i.e. the management approach actually adopted, and on the other hand the societal value system represented by formal law, widespread opinions, convictions etc. But values are also influenced by the long lasting use of special professional methods: The ongoing use of e.g. HIPO-diagrams (which are oriented to information flow aspects in programmes) leads to a different understanding of the design process than the use e.g. of interaction nets (which focus on the information exchange between user and system instead).

That there exist differences in design values, has been shown by Kumar and Bjørn-Andersen (1990) in a cross-cultural study of the orientations of information system designers in Canada and Denmark. Firstly, it could be shown (cf. Tab. 1) that for both groups of system designers technical values have a much stronger influence on their design decisions than socio-political values.[2] Secondly, regarding the material presented it becomes obvious that the national background of the designers plays an important role in their design orientation: Canadian information system designers showed a significantly stronger orientation towards technical values and a weaker orientation with respect to socio-political values than the Danish designers. Whereas the Danish designers show an almost balanced relation between the value categories, the value scores of the Canadians are highly divergent. The interpretation of Kumar

---

2 Examples of the value categories used in the cited study are: "Technical values": reliability of the system, compatibility with interfacing systems, ease of producing and maintaining documention etc. "Economic values" are: operating costs of the system, development project on schedule, manpower required for project etc. "Socio-political values": user's autonomy in planning and performing the task, job induced mental stress on user, participation of users etc.

and Bjørn-Andersen assumes that the higher scores concerning technical values in the Canadian case lead to an emphasis on *efficiency* (cost, manpower usage etc.) whereas the more balanced value profile of the Danish results is stressing the *effectiveness* of a system (management of organisational resources etc.). Assuming that, because of the growing complexity of future information systems, effectiveness will be more important than efficiency, the more balanced Danish value profile will not only lead to a higher level of humancentredness but also to a higher level of competitiveness.

Average Behavioral Relevance Scores [3]	Technical values	Economic values	Socio-political values
Canadian designers	52,4	48,7	30,5
Danish designers	47,1	46,2	40,3

Tab. 1: Value profiles of Canadian and Danish information system designers (Kumar/Bjørn-Andersen 1990, p. 533)

Brödner (1990) worked out that the design of technology can be seen from the viewpoint of two very different basic approaches: 1. the *technocentric* design culture, 2. the *anthropocentric* design culture. Whereas the technocentric approach focuses on the high productive potential of computerised technology and seeks to reach the state of full automation with no personnel required, the anthropocentric approach emphasises the central role of the worker's knowledge, skill and experience in organising and performing the production process, being supported (but not replaced) by computerised machines. Brödner showed that in the case of factory and work design in the machine-building industry, the technocentric route is still dominating. But he also put forward strong arguments that the technocentric way in the long term will not be successful, neither in social nor in economic terms.

The situation in other engineering disciplines, including informatics, is very similar to that described by Brödner. Engineering has its own historical "culture of rationalism". The engineering disciplines are bound up in the tradition of the mathematics and natural sciences of the "Age of Enlightenment" and as such grew up in opposition to the largely irrational explanations of the social, philosophical and religious systems of those times. Then, engineering gained its strength just by concentrating on the technical core of a construction problem. The discipline defined

3 "The BRS [Behavioral Relevance Scores] for a value dimension is defined as the percentage of the total group for whom the particular value dimension is an operative value. Accordingly, a high BRS for a value dimension for a group indicates that, in that group, there is a high likelihood this value would be translated into actual behavior." (Kumar/Bjørn-Andersen 1990, p. 531)

itself by focusing on this basis approach, and implanted it as a deep conviction into the educational training of technicians and engineers over a hundred years.

The *technocentric design culture* in software development appears in three forms:

a) *Neglecting* the user (or the affected) and his human characteristics, interests etc. In this approach only the technical part of the system is considered to be relevant for software design. Though it is admitted that the user is part of the system, he nevertheless plays no role in the programming process itself. The analysis of tasks and user requirements from this point of view has to be conducted within the application area, resulting in a document containing a precise and detailed *formal specification* which can be seen as a "software treaty" between programming and application department. The goal of software engineering is to translate this formal specification directly (automatically) into programming code the correctness of which can be logically proven.

b) *Reducing* the user to a data entering component. In this approach the user is actually included in the software development process but his function is reduced as far as technically possible. System designers try to fully automate the task. The main goals are "completeness of programme code", the reduction of user interference and the prevention of any user mistakes (foolproof systems). The remaining tasks for the user (data entry, bridging system gaps etc.) are mostly tedious, cause heavy mental load and de-skill the worker.

c) *Simulating* the user in terms of an (intelligent) machine. The third form of the technocentric approach fully involves the user into the system as an equal part(ner) of function and communication. But, to operate in a homogeneous environment the function, knowledge and performance conditions of the users are described in terms of computer systems (cognitive science approach). This method neglects most of the important human factors in computer-assisted working environments (informal communication, intuition in problem solving, the role of work experience etc.) and alleges a symmetric relationship between user and computer though in fact it should be a question of a master-slave relation.

But there also exists an *anthropocentric design culture* in software development, even if this approach is only supported by a minority of designers:

The main difference of this approach compared with the technocentric route is the acceptance of the leading role of human being in the socio-technical system. The anthropocentric design orientation is lead by the idea that the computer should be used as a "sophisticated tool" which offers a high degree of work flexibility to the user. Putting the user into the centre of the computer supported work process is based on the idea of improving the productivity of the user by augmenting his intellect, knowledge and experience instead of trying to replace these human abilities by a machine. Following this basic philosophy Volpert formulated a number of rules from an occupational psychologist's point of view which should be taken into consideration by software designers. These "human criteria" (cf. Tab. 2) focus on a self-determined active processing of the given work task.

The application of this anthropocentric design orientation is not only a question of a "humanisation of working life" but also a matter of productivity improvement. Therefore the systems designer has to reflect on the fact that the idea of a "workerless factory (or office)" can not be the aim within an industrial system which just gains its competitive edge and thereby its profit mostly by "implementing" high qualifications into its products.

Human criteria: Elements of a human-centred design culture for computerised work places
1 Work tasks must allow for a broad scope of action.
2 A wide temporal scope must be provided to perform a task
3 Work tasks must provide facilities for developing a personal approach to grasping and mastering demands in terms of structurability (e.g. enhancement of system transparency, offering adaptation tools).
4 Work tasks must avoid objective hindrances to the work activity (e.g. organisational or technical deficiencies and faults).
5 Work tasks require provision for sufficient and varied bodily activity, they must call for the use of a wide variety of sensory capacities, and they must provide for the concrete handling of real objects and ensure a direct relation to social conditions.
6 While the basic structure of the tasks remain the same, they must provide for a variety of different implementation conditions.
7 Working conditions must provide for and encourage co-operation and direct interpersonal contact.

Tab. 2: Human criteria (Volpert 1992)

2.2 Example: The adaptivity-adaptability-controversy

The discussion about designers' orientations shall now be illustrated by a concrete question in the context of human-computer interaction: The question is whether a system should be *adaptive* in terms of the user or whether it should be *adaptable*. There is, as is explained later, a huge difference between these two concepts which mainly refers to the question: Who takes the *initiative* in the human-computer interaction? The controversy is strongly influenced by the traditional human factors approach and the designers' orientation linked to it. Therefore the first step should be to highlight the differences between the physiological and the cognitive ergonomic concept.

The classical (physiological) ergonomic approach demands the adaptation of the machinery to the abilities and properties of the human being. This is due to the fact that the physical abilities of people are fixed within certain limits. In contrast to this, modern (cognitive) ergonomics require a double-sided mutual adaptation: In the case of fixed cognitive abilities (e.g. the limited capacity of the short term memory) an adaptation of the computer system to the employee; in the case of formable ("plastic") cognitive abilities the adaptation of the working person to the computer ("learning"). Both kinds of adaptations aim to meet the requirements of dynamic system behaviour: a) The growing knowledge of the user about the system has to be in accordance with a growing range of functions offered by the system and an adequate interaction facility. b) The different situations a user can be involved in (stress, rare usage of the system etc.) also necessitates the possibility of system adaptation at any time.

The question of how this ability to adapt should be realised, has been answered in considerably different ways in the field of software-ergonomics: Should designers focus on the concept of "adaptivity" or "adaptability" (cf. Friedrich 1990)? As we will show, the answer to this question is strongly related to the industrial cultural background of the respective designer.

Adaptivity is the property of a system to adapt itself automatically to the changing behaviour of the user. Examples of adaptivity features are the automatic correction of user mistakes without user-feedback or the automatic shift of system performance parameters (e.g. the speed of the mouse movement or the level of usable subsets of the command language) because of a certain behaviour of the user at any given time (being heavily burdened for a moment or so). Adaptive systems usually work on the basis of an internal artificial intelligence (AI) model of the user which has been generated by an observation, classification and storage of user behaviour in different situations (learning systems). The system then reacts, following given criteria, to an actual or even expected behaviour, in an "intelligent" way. Besides privacy problems, the main critique that has been made against this approach of adaptivity aims at the fact that the user is condemned to passivity and his opportunity to learn is significantly reduced.

In contrast to this the *adaptability* of a system offers the opportunity to the user to tailor the system to his current competence and working situation by himself, using it like a tool box. By this "concept of individualisation" (Ackermann, Ulich 1987) the user becomes the active element of the man-machine-interaction. He alone decides the (variable) strength and requirements of the system, the kind of programmes to use and the order of carrying it out. Such an approach gives the user a diversity of possibilities in acting and deciding, in combining planning and doing, and requires a high degree of reflection upon his own work. Last but not least the approach of adaptability is through necessity founded on the concept of user participation, not only during the use of a system but also (and even more important) in the earlier phase of system development.

There already exist prototypes realising the principles of adaptability. The precondition of a user-adaptable system is a high degree of system transparency. With the "sites, modes and trails" model Nievergelt (1980) has presented a famous example of a transparent system on the basis of the operating system XS 2. The user can put questions to the system like "Where am I?", "What can I do here?", "Where can I go?", "How can I get there?" etc. Answering these questions requires a highly sophisticated system, usually on an AI basis. (This shows once again, that there are social choices in developing and using advanced technologies such as AI.) The most advanced adaptable systems are the so-called qualification-promoting systems. They "encourage" the user to proceed from simple to sophisticated (and by this more powerful and effective) system variations by system supported learning while working (cf. Palme 1983). They offer "navigation instruments" to discover the system. The user adjusts the complexity of the system functions by himself, e.g. the different modes of interacting by menus, by object manipulation or by using a command language (at different subset levels). At the same time these systems are fault-tolerant in the sense that the user can perform an UNDO- or a REDO-command to avoid frustrating situations.

Summarising the discussion, one can say that there exists a strong competition between these two paradigms within computer science the outcome of which has not yet been decided. These paradigms are highly correlated to the national and cultural environments the software designers are working in. The adaptability approach is favoured by continental European and especially Scandinavian software designers (cf. Nygaard 1986; Bjerknes, Ehn, Kyng 1987; Ehn 1988; Greenbaum, Kyng 1991). Working life in these countries is highly regulated by legislation in the working area, compulsory ergonomic standards and collective agreements. Trade unions and works councils are strongly engaged in pushing through these regulations at company level. In the continual conflicts about the realisation of these ergonomic demands software designers are evolving a broad awareness of the necessity of human centred design. It may be too that the special structure of the work force in these countries is strengthening the adaptability approach: There is a broad group of medium qualified workers, based on a traditional apprenticeship system, which are used to adjusting their working tools to their tasks. On the other side, there are countries as e.g. the U.S. with a liberal, only weakly regulated labour market. Software designers in these countries are oriented more towards functionality and productivity than to human factors. Besides this, the work force in these countries is more strongly stratified with respect to qualification: There exists a larger number of higher educated people (college level) but also a greater number of unskilled factory and office workers. Computerised automation affects above all this second group of unskilled workers. For them it may be difficult, at least without further training, to adapt the system to their own work. So software designers prefer an automatic adaptation mechanism (adaptivity).

What becomes obvious from the presented example so far, is that the designers' orientations towards adaptability or adaptivity in software development is highly correlated to factors such as industrial relations, qualification structure, work legislation system etc. of the respective culture, country or region. This complex framework of interdependent factors has to be differentiated by further research.

3 Industrial Culture II: The Industry of Software Production

The software development process is not only determined by the mental orientations of designers but also by the organisational concept of the management. From a management point of view the generation of software is characterised by a number of deficiencies: lack of controllability of the designers' work, low degree of productivity compared with the production area, exceeding of time limits and budgets, dependency on a highly qualified labour force which is in short supply etc. The reason for these deficiencies is seen in a lack of standardised and structured engineering methods. Therefore there is a double tendency towards standardisation: Standardising the process of software production on the one hand and standardising the products, i.e. the written programmes, on the other. Twenty years ago software development was still described as the "art of computer programming" (Knuth). Every programme had its own "hand writing". The way of structuring the procedures and of optimising the code depended mainly upon the qualification, experience and habits of the programmer.

The first organisation which experienced a problem with this more or less intuitive style of programming was the military. This bottom up approach did not fit into the hierarchical well structured military organisation. So the first international conference on "Software Engineering" was organised by NATO in 1968. The task then was to elaborate and define the principles of "top down approach", "structured programming", "modularisation", "information hiding" etc. Civilian edp has followed this way "from art to engineering". It has enabled enterprises to strengthen the division of labour among edp personnel (cf. e.g. the IBM concept of "chief programmer teams") and to introduce computerised methods to facilitate and (partly) automate the software development process. It is small wonder that the construction and combination of software development tools like compilers, debuggers and software libraries tend to become one powerful, complex and uniform "software development machine". At present these attempts are discussed using the term "CASE tools" which means "computer aided software engineering" tools. Even if these programming environments are labelled as "tools", in their dominant mode of application they in fact represent a "machine perspective" in the sense of replacing and deskilling the work of programming. This machine perspective leads to the next step in the contemporary discussion of software engineering: the concept of a *software factory* (cf. e.g. Evans 1989). By this concept the basic principles of material (mass) production as has been realised in the traditional factory are to be transferred to the area of "cognitive production" of programmes. Among others these principles are (in brackets the edp related translations): application of machinery to facilitate work (in edp: application of CASE tools), decomposition of tasks into simply performable sub tasks (-modularisation), combining the results of the sub tasks by using assembly lines (module linking, programme integration), specialisation of the work force aiming at a higher degree of division of labour and a lower demand for qualification (system analysts, programmers, coders), separation between planning and performing of tasks (chief programmer team), introduction of standardised methods and means of production (structured programming, reusability of software components), controllability of work performance and product quality (computerised project tracing, in-line programme quality assurance). The concept aims at the realisation of the CIM philosophy in software production.[4]

Summarising, one can say that the mainstream discussion in software engineering is not only characterised by the tendency towards standardisation but also by the attempt to implement development methods which have proved to be very successful in the area of material production. That means to focus on a machine and factory perspective of software development. As we have already seen in the case of designers' orientation, the attempts to implement the software factory concept also depend on the dominant industrial cultural factors in the various countries. So, this approach is stronger represented in those countries which have a powerful sector of mass production (like the U.S. and Japan) and less in countries with a broader share of specialised small and medium scale production (like for example in Germany).

Thus it is not surprising that again in northern and central Europe there is a discussion about alternatives with respect to the software factory approach. In this situation it seems very likely to suggest the model of a "programming workshop" (cf. Budde, Züllighoven 1992). The work in a programming workshop is characterised by the production of specialised ("consumer tailorable") software. This kind of software is located between (mass oriented) standard software, like text processing systems or spread sheet programmes, and individual "one-off" programmes. Workshop produced software fulfils the needs of small and medium sized user groups, with special application problems and high demands with respect to flexibility in programme use. To produce such a kind of software calls for highly qualified system designers who are able not only to be in control of the technical process but also to communicate with a wide range of different customers and gain an understanding of their specific needs. It is obvious that this kind of software production needs other methods of software engineering than the mass production oriented approach. Working together with customers and users requires qualifications in "user centred communication", methods for "continual user participation", tools to realise "rapid prototyping" etc.

The discussion in chapter 2 about the orientation of systems designers can be repeated here with respect to the level of management: The *design philosophy* (as for example reflected in the adaptivity-adaptability debate) was significantly determined by the "user model" of the designer. This user model is formed by the prevailing industrial cultural factors of the working situation and their subjective perception by the designer. The *management philosophy* of software production is, in a very similar way, determined by the "designer model" of management.

In trying to offer a preliminary conclusion considering the relationship between

---

4 CIM means "Computer integrated manufacturing" and is often used as a metaphor for the concept of a fully automated factory. Besides the question if a CIM system can be designed in a human centred way (cf. Corbett, Rasmussen, Rauner 1991), because of its inherent complexity and formal rationalistic approach there is a reasonable doubt if this concept will at all function in its original domain, namely the material production area.

industrial cultural factors and software production, one can say that there does not exist a linear deduction mechanism: Industrial cultural factors are not only the *basis* of different software design and production approaches but also the *result* of the application of the respective design philosophies. In software production there seems to exist a mutual promotion of industrial cultural factors and design strategies. These findings which are based on an extensive study of literature on the one hand and a number of unstructured interviews in software development practice on the other have to be proved empirically in the next step of our research project.

References

Ackermann, D.; Ulich, E.: The chances of individualization in human-computer interaction and its consequences, in: Frese, M.; Ulich, E.; Dzida, W. (eds.): Psychological issues of human-computer interaction in the work place. Amsterdam, New York, Oxford, Tokyo 1987, 131-145

Bjerknes, G.; Ehn, P.; Kyng, M. (eds.): Computers and democracy. Aldershot etc. 1987

Brödner, P.: The shape of future Technology. The anthropocentric alternative. London 1990

Budde, R.; Züllighoven, H.: Software tools in a programming workshop, in: Floyd, C.; Züllighoven, H.; Budde, R.; Keil-Slavik, R. (eds.): Software development and reality construction. Berlin etc. 1992, 252-268

Corbett, J. M.; Rasmussen, L. B.; Rauner, F.: Crossing the border. The social and engineering design of computer integrated manufacturing systems. London etc. 1991

Dagwell, R.; Weber, R.: Systems designer's user models: A comparative study and methodological critique, in: Communications of the ACM, vol. 26 (1983) no 11, 987-997

Ehn, P.: Work-oriented design of computer artifacts. Stockholm 1988

Evans, M. W.: The software factory. New York etc. 1989

Friedrich, J.: Adaptivität und Adaptierbarkeit informationstechnischer Systeme in der Arbeitswelt - zur Sozialverträglichkeit zweier Paradigmen, in: Reuter, A. (ed.): GI - 20. Jahrestagung I. Berlin etc. 1990, 178-191

Greenbaum, J.; Kyng, M. (eds.): Design at work. Cooperative design of computer systems. Hillsdale 1991

Hedberg, B.; Mumford, E.: The design of computer systems. Man's vision of man as an integral part of the system design process, in: Mumford, E.; Sackman, H. (eds.): Human choice and computers. Amsterdam 1974, 31-59

Hildebrandt, E.; Seltz, R.: Wandel betrieblicher Sozialverfassung durch systemische

Kontrolle? Berlin 1989

Kumar, K.; Bjørn-Andersen, N.: A cross-cultural comparison of IS designer values, in: Communications of the ACM, vol. 33 (1990) no 5, 528-538

Nievergelt, J.; Weydert, J.: Sites, modes, and trails: Telling the user of an interactive system where he is, what he can do, and how to get places, in: Guedj et al. (eds.): Methodology of interaction. Amsterdam 1980, 327-338

Nygaard, K.: Program development as a social activity, in: Kugler, H.-J. (ed.): Information processing 86. Amsterdam, New York, Oxford, Tokyo 1986, 189-198

Palme, J.: A human-computer interface encouraging user growth, in: Sime, M. E.; Coombs, M. J. (eds.): Designing for human-computer communication. London 1983, 139-156

Rauner, F.; Ruth, K.: Comparing industrial cultural determinants of CNC-machine-tool building in the United States of America and in the Federal Republic of Germany. Bremen, Dec. 1988 (paper)

Rauner, F; Ruth, K.: Perspectives of research in 'industrial culture', in: Proc. of the 1. CAPIRN conference (11.-18.12.1989). Santa Cruz 1989

Volpert, W.: Work design for human development, in: Floyd, C.; Züllighoven, H.; Budde, R.; Keil-Slavik, R. (eds.): Software development and reality construction. Berlin etc. 1992, 336-348

Industrial Culture - An Action-Oriented View at Innovation and Production

by Klaus Ruth

1 Introduction

Recent research on innovations and production has, as a rule, very successfully opened itself to new influences and metaphors, as long as it strove to break out of the corset of reductionist economic explanatory logic. The range, especially of innovation approaches, varies from political science oriented institutional concepts and "culturally enriched" social science technology research to organizational theory models. From a superficial point of view the industrial culture approach developed since 1987 might also be misinterpreted as culture-sociological technology research, although the label industrial culture has more to do with a comprehensive approach for interpreting the social organization of innovation and production. In this way emphasis is placed on the specific characteristics of innovation and production processes in different national (in our terms: industrial culture) contexts and, above all, an attempt is made to explain differences in the development and application of programmable automation (PA) techniques as a result of different industrial cultures.

Up to now research interest (as reflected by relevant research projects of the CAPIRN research network currently in progress) has concentrated on the development and application of technology and thus merely looks at partial aspects of the entire complex of production and innovation. Nevertheless, I would like to assert that the approach proposed here can be attributed general validity for production as well as for technical and nontechnical innovation processes. I realize that this assertion cannot be verified by this paper alone, just for the reason that the empirical basis used up to now with PA techniques is too narrow for a general industrial culture theory of production and innovation. Interested readers are referred to other contributions in this publication and to upcoming work.

This paper will attempt to further develop the industrial culture research approach theoretically and thus, at the same time, to make a contribution to systematizing and "codifying" already available, particular cognitive building blocks. A major function of this work is to bring together micro- and macrotheoretical components; at the microlevel industrial culture dimensions have long been developed and made operational in concrete empirical research instruments while at the macrolevel there are elaborated hypotheses on industrial-cultural processes of overlapping/acculturation and on convergence/divergence cycles in the field of development of PA techniques (Rauner, Ruth 1990). The differing levels are analytically linked by virtue of an action theory concept whose central term, action, functions as a means of transmission between industrial culture, on the one hand, and innovation/production processes, on the other hand.

2 Technology and Industrial Culture - Two Incompatible Concepts?

In the beginning there was uneasiness: triggered by developments on the world market for metal-cutting machine tools in the second half of the 80s, an "explanatory crisis" suddenly broke out. What was the cause of the rapid decline of the U.S. machine tool industry and the unparalleled rise of the Japanese in this field as well as for the consolation of the market position of traditionally strong European manufacturing nations, such as Germany, Italy and Switzerland? Were these developments related to the immanent momentum of CNC technology or could they be explained with economic factors and arguments connected with sectoral structure? Relevant research focused on the search for explanations for the decline of the U.S. machine tool production industry because they would then furnish arguments for the success of the other manufacturing nations. Tax legislation and specific features of demand (hypercyclicity of demand for machine tools, overdependence on military complex, etc.) in addition to exchange rate arguments were particularly attractive explanations for this zero sum game of lost and gained world market shares. Very rarely were weak innovative capabilities of the sector analyzed as the cause (MIT Commission 1989).

The essential, and justifiably questionable, assumptions on which these approaches are based fail to see any differences in technological artifacts themselves, which would then require additional explanation. Moreover, they completely overlook the idea that possible differences in technology might be one cause of competitive advantage. Yet in studies conducted by Kuby (1980) and Radkau (1989) dealing with the pre-NC period, by Noble (1984) for the NC phase as well as others for the CNC era, a large body of evidence has been compiled that strongly supports the notion of differing "brands" of technology. Our first studies, which focused in particular on man-machine interfaces, confirm this suspicion (Rauner, Ruth 1989; as well as Hirsch-Kreinsen 1989).

Proceeding from this "initial suspicion", the attention turned to the social forces outside the technological and economic fields that are responsible for certain developments in CNC technology. The key idea for the overall approach to this question soon took shape in a concept labeled industrial culture. This approach took into account factors that to date had been used very rarely in connection with the development of technology and its application; and as a consequence the reservations and misunderstandings voiced by researchers in this field were considerable.

The concept of industrial culture - to judge by past discussions - has been the source of frequent misunderstanding. The most common one that spontaneously arises is based on the juxtaposition of the two words. What, one might ask, does industry have to do with culture, that noble sphere dedicated to bourgeoise ideals? Does not such an approach fall under the auspices of an industrial archeology project that investigates the degree to which industrial-technological artifacts are culture-bound, or is this concept ultimately expected to demonstrate the increasing industrialization of culture?

All of these misunderstandings could be cleared up quite quickly, not least of all because they were of a superficial, terminological nature. A deeper yet by no means less problematical question in terms of terminology and categories lies in establishing a definition for industy, technology and culture. The industrial culture approach does not regard technology and culture as two opposing concepts between which a strict distinction has to be made, as in the hardware/software dichotomy for example, but rather as two interacting quantities. Culture can manifest itself in the material world, while technology has always had an intellectual component (cognitive as well as cultural), on one hand, and, on the other hand, it affects culture through its material artifacts as well as through its representation in human thought. These two aspects have also been dealt with by technology research as carried out by traditional sociology. This field has seen the emergence of two basic lines of research:

a. The first line of thought aims at revealing technology's social effects, while
b. the second concentrates on ascertaining the social influences on technological development.

The first line of research regards technology as an independent variable, assuming that technological development proceeds largely in the absence of social forces. The concept of technology used here most closely resembles the idea of a black box (see, among others, Rosenberg 1982), which is not subject to social forces but which essentially follows an inherent techno-logic (i. e. an inherent momentum/logic of technology which is unaffected by social forces)[1]. At most, economic influences are occasionally taken into account, but even then their function is generally one of inherent necessity. Typically, the questions posed by the first category of (sociological) research deal only with finished artifacts, thus concentrating more heavily on the effects and consequences of (actual) technology.

The second line of research regards technology as the result of a process that takes place in the interactive field of social (and cultural) forces (see Mumford 1934; and Ullrich 1979). Approaches of this type occasionally run the danger of placing too much (or exclusive) emphasis on social factors, while sorely neglecting the inherent logic of technological development.

While scientific interest in the first type of approach seems to be waning, recent and predominantly economic debates concerning models of innovation have increasingly focused on the second approach outlined above - but not, in order to avoid any misunderstandings, in the extreme form of "social determinism".[2]

For the industrial culture approach the significance of these two research traditions lies in the fact that it incorporates the merits of both concepts and unites them

[1]One of the most famous advocate of the therefrom arising technocracy-theorem is Schelsky (1965).

[2]See the following section of this essay entitled Current Debates.

dialectically. Accordingly, this approach is expected to take into account technology's ability to shape culture on one hand and at the same time examine social and cultural influences on technology. Technology as a factor in shaping culture or as a decisive determinant in influencing our scientific and technological civilization has long been a topic of debates concerning so-called everyday technologies.[3] Technological artifacts such as automobiles or walkmans have been attributed forces that have a harmonizing effect on civilization as well as ones that drastically reshape and differentiate it (Rammert 1988, p. 186ff). Here the industrial culture research program might include examples taken from the industrial sphere, since certain manufacturing and organizational technologies, such as Ford's assembly line, obviously have very far-reaching effects on various aspects of industrial culture (e.g. management-labour relations within the company).[4] At the same time industrial cultures having such a predisposition also influence industrial technologies, their development and especially their application in a specific manner. Such interrelationships have been sufficiently documented by studies dealing with turbine development (Layton 1978), electricity supply systems (Hughes 1983) as well as our own studies concerning NC/CNC machine tools (Rauner/Ruth 1989 and 1990; see also Noble 1984, Hirsch-Kreinsen 1989).

Up to now the industrial culture approach has primarily aimed at investigating the effects of industrial culture on technology, its development and application. In keeping with the dual concept of technology and culture, especially with respect to its claim of presenting a dialectical approach, it is also important to examine more closely the effects of technology on industrial culture. An assessment of the connecting lines and mediatory steps that exist between culture and technology as well as between technology and industrial culture is essential to both of these topics, or lines, of inquiry. This can be accomplished by developing our concept of industrial culture along lines that are both subject-oriented and action-oriented. The main focus of this approach is therefore directed toward actions as well as agents of action, for technology can be studied as a given and completed quantity, in which case it would be the result of actions carried out by individuals having a particular orientation. But on the other hand, if technology itself is seen as the performance of an action, the question as to the agents of this action and their motives suggests itself even more. An action-oriented theory is thus an impressive way of highlighting the connection between (these two views of) technology and culture.

In the following the concept of industrial culture will therefore be developed as an

---

[3]At this point I would briefly like to point out the distinction between culture and civilization. civilization, in the sense of a scientific and technological civilization, refers more to global unifying tendencies, while culture, as I understand the concept, concentrates on global processes of differentiation.

[4]It would be interesting to investigate the degree to which a centralized technology tends to promote a conflict-oriented form of workplace social relations while decentralized technologies tend to be accompanied by more cooperative forms - or vice versa.

action-orientated theory. This will not only help to clarify and reappraise the concept's theoretical points, but through the use of an action-orientated means of interpretation it will also remove the concept of industrial culture from a highly theoretical plane to one having a certain degree of complexity and operability. The actual advantages this method has for the analysis of industrial culture and for the interpretation of innovation and manufacturing processes will be discussed in a second step. This is particularly important, for it means that the empirically well-documented field of "CNC machine tool development and application" can be extended by induction to general processes of innovation and production.

I would first like to examine briefly the current debates concerning *national innovation systems and economic culture*, since I feel that a thorough examination of the similarities and differences in these approaches will highlight the special nature of the industrial culture approach.

3 Current Debates

As a direct result of the dissatisfaction with standard theories of economic growth, innovation theory of the last few years has taken a new direction in its analysis and research. Generally referred to by the catchword National Systems of Innovation (NSI), it aims at ascertaining the sources and processes of innovation (Dosi et al. 1988). Its analytical focus centers on "[t]he network of institutions in the public and private sectors whose activities and interactions initiate, import, modify and diffuse new technologies..." (Freeman 1987, 1). This general definition assigns an important role to the processes of innovation, long regarded by economic theory as an ambiguous residual quantity, in the study of economic development and industrial competitiveness. In the process this means that **institutions** assume a decisive function. The national climate for innovation is influenced not only by institutions directly involved in R&D activities, but also by those related to national organization and resource management. Accordingly, government and government policy would be the relevant actors on the national level alongside universities and public research institutions. Furthermore, Dosi and Orsenigo also regard as institutions "... all forms of organisations, conventions and repeated and established behaviours which are not directly mediated through the market" (Dosi, Orsenigo 1988, 19). This definition would include - in my interpretation - most notably informal user-producer interactions, which have proved to be a relatively significant structural characteristic of processes of innovation and diffusion within national frameworks (Lundvall 1988, 360 f; von Hippel 1988).

Furthermore, the existence of national innovation systems underscores the importance of strong coherent forces within national economies, for it is precisely these structures that generate the uniqueness of NSIs and thus characterize national differences. According to NSI representatives, differences between individual NSIs can be found in a variety of social spheres. Differences in the "inherent factors of production, cultural and historical factors, geopolitical variation, and general economic orientation" are named in particular (Crow, Bozeman 1991, 2). With these factors the NIS approach resembles the industrial culture concept. The most glaring difference between NSI approaches and the industrial culture concept can be found in the research objectives. Based on Schumpeter traditions, NSI theories attempt to place the emphasis of economic analyses on technical and institutional change (instead of neglecting them to a greater or lesser degree as residual factors) and creating links to micro- and macroeconomic adjustment and transformation processes. The industrial culture approach does, in part, look at identical variables (e.g. social institutions), but focuses, in contrast, on an analysis of technology development and application processes as well as on technology itself. Whereas in the latter case only very indirect statements are made about technological competitiveness and structural change, the epistemological focus of NSI approaches is on the dynamics of economic growth, competitiveness, business cycles and the problems of economic equilibrium.

In addition to the NSI debate, there is another complex of approaches, generally grouped under the label **economic culture**, which are just as relevant to the industrial culture approach. The concept of *economic culture* as such has not yet been outlined as a coherent approach (or paradigm), but major issues have been developed in relatively clear terms by Hofstede (1980), Berger (1987) who coined the term, and Clegg et al. (1990). The economic culture approaches presuppose differences in the cultural configurations of national economies. The constitutive assumption here is that differences in economic cultures are reflected in the economic performance of national and regional economies.

In their fundamental work Bond and Hofstede (1990) investigate the link between culture and economy by distinguishing four main cultural dimensions:"...power distance, individualism, masculinity versus femininity, and uncertainty avoidance" (Clegg, Redding 1990, 4) and then transposing these to standardized empirical tools. Their comparative studies failed to verify an explicit (causal) connection between economic culture and economic performance (in terms of per capita GNP, etc.). This was not the case when removing uncertainty avoidance and introducing a fourth cultural dimension of **Confucian dynamism** which varies from "a dynamic future-oriented" towards a "static tradition-oriented mentality" (Clegg, Redding 1990, 5). Bond and Hofstede conclude that there is a **Confucian cultural ethic** (in analogy to the Protestant Ethic) which is highly influential in the Asian context, This model provides a cultural variable that actually does correlate empirically with economic success/growth and is particularly well-suited for explaining the economic success of developing countries in Southeast Asia.

Whereas the approach sketched above focuses on 'macroeconomic culture' and on the 'micro-cultural level' at the same time, in the way that the sediments of culture in the minds of individuals are explored by using the concept of cultural dimensions, there are others concentrating on the meso-cultural level. These approaches concentrate on companies/organizations and/or bureaucracy/management to explain the dynamics of economic success of countries or enterprises. Far from being enterprise or corporate culture economic culture approaches focus on normative and institutional frameworks influencing the (economic) action of companies or organizations. In search for an explanation of the success of Japanese (and other Asian) organizations Clegg et al. develop the concept of **post-Confucianism** which is characterized by a "stress on familialism, collectivism defined in terms of the family, and a meritocratic stress on education as the means to collectively consolidate family wealth" (Clegg et al. 1990, 38). To avoide culturalism Clegg et al. devise an institutional framework within which social, political and cultural factors are represented as the key-elements of economic culture.

Indeed, vicarious for many economic culture approaches Clegg et al. can give plausible explanations for the correlation between culture and economic success but one fundamental difficulty is their lack in explaining the nature of the connection and the levels of mediation between culture and economy. It is especially this interaction between (industrial) culture, on the one hand, and innovation and production, on the other hand I want to focus on in the following paragraphs.

4 Action-the Missing Link

In defining the concept of industrial culture, it is meaningful to do so from the perspective of the projected area of study. Innovation and production are the fields of action within which the significance of the determinants of industrial culture would have to be ascertained.

First of all, I want to look at innovation and production as processes, in the course of which technical artifacts are developed or applied. By explicitly including the processes of the creation and application of technology, I adapt the concept of *real technology* (in German language: Realtechnik), which Gottl-Ottlilienfeld developed in the 1920's (Gottl-Ottlilienfeld 1914). In his concept the term real technology is distinguished from individual, social and intellectual technology. His distinction basically consists of the differentiation between object-related technologies (real technology) and humanistic technologies (or immaterial technologies), under which social, individual and intellectual technologies fall. He concedes that real technology also contains socially, individually and intellectually technical components.

Gottl-Ottlilienfeld regards technology in the objective sense as the entirety of processes and aids of nature-dominating action within a certain area of human activity (Gottl-Ottlilienfeld 1914, 206). Real technology essentially comprises

a. material artifacts,
b. the sum of all actions from which these artifacts stem and
c. the entirety of all actions, whose focus is on the application of these artifacts.

The concept of *real technology* is thus excellently suited to linking an action-oriented industrial culture concept with technology, development of technology and application of technology.

The reason for performing a reduction to material artifacts at this point is to make it easier to deal with the concept. At the same time, I realize that a broader concept of technology in the sense indicated above, which also includes soft technologies, i.e. organizational techniques, etc., is to be used as the basis in a subsequent step. The restriction of the technology concept to material artifacts is to be understood from a project history point of view, it stems from the original machine-based project topic.

First of all, I wish to present an initial approach to the content of industrial culture by making reference to *classical* concepts of culture and action. M. Weber defines cultural value spheres as a stock of shared values and implicit rules out of which individuals and groups obtain their patterns of action and interpretation (Weber 1972).

According to Weber, it is the stock of shared values which makes meaningful action and non-contingent interpretation of it possible in the first place. In this construct we can already make out the interface between culture and action which is relevant for the concept of industrial culture. In his basic sociological concepts, Weber makes a distinction between various ideal types of action, i.e. **reactive, meaningful** and **social action**. Reactive action is to be regarded as traditional behavior not connected with a subjective meaning, meaningful action can be more rationally or emotionally influenced. Weber, however, sees the central subject matter of sociology in social action and this type of action is also of special importance for our concerns. Social action can be defined in four ways. It can follow a calculated, rational purpose (in German language: Zweckrationalität), it can be oriented to rational values (in German language: Wertrationalität), it can follow emotional motives or be traditionally determined (Weber 1972, 12ff)[5]. It is important to point out here that Weber devises these action orientations as ideal types and, therefore, reality can only approximate them. In the same manner, the strict separation of the individual action orientations is to be interpreted as an analytical structural aid-an exclusivity which only sees one or only the other type of orientation as a determinant of action does not exist in reality; instead one must always assume the existence of mixed forms.

The significance of the theoretical recourse to Weber's concept of action for my primary area of interest, innovative action[6], derives from the fact that with the four orientations of action a scheme is available which, going beyond rational decision-making models, regards values and traditions, i.e. culture as well as emotional aspects, as relevant to action.

The linking of cultural with action and interpretation processes in Weber's concept is to be seen as an important step towards developing the industrial culture approach in that here actions are subject to a bundle of orientations that can be specifically configured in different contexts in each case. This lays the foundation for a culturally (in our terminology: according to industrial culture) varying action orientation matrix, by means of which the different technology developments can be explained for the field of innovation. That leaves the question open as to how to develop the concept of industrial culture and categorically distinguish it from the concept of culture. I would like to quote T. Parsons' theory of action systems in this connection. It seems as if we had to drag a great deal of ballast with us on our way in the opposite direction on this

---

[5]Weber's research interest concentrates on rationality and not primarily on culture, but since it includes culture, the concept can also be used in our sense.

[6]Production action has equal relevance, but is left out at this point - solely because of the brevity of this paper.

busy road. However, the reason that I only refer to one of the core areas of Parsons' theoretical edifice, i.e. the action system theory, is because this very portion of the entire structure is tailor-made for our specific question under study.

In Parsons' variant of the action system theory the (Weber) problem of rationality is widely left out. Instead, there is a focus on the cultural aspects of action systems, i.e. the processes of transmission between culture and actions. Because the steps of transmission are to be regarded as a central problem of the concept of industrial culture, I wish to go into this point somewhat.

Parsons follows Weber's concept of culture as a value consensus, but emphasizes to a greater degree the significance for and transmission levels between action/interaction systems and those performing actions. Regarding formative aspects, Parsons distinguishes between three configurations of action systems:[7]

a. personality;
b. social systems;
c. culture.

Parsons defines personality "as the organized system of the orientation and motivation of action of one individual actor (Parsons, et al. 1962, 7)". Parsons conceives of social systems as the interaction of a great number of actors in common situations (social interaction). Here it is the interaction process itself that is the focus of study. Personality and social systems are closely connected to one another, but neither are they identical, nor can one be explained on the basis of the other - accordingly, the social system may not be misunderstood as a large number of personality systems.

Finally, cultural systems, as still has to be shown, are the most significant for our approach. According to Parsons, the former are primarily characterized by the fact that they are not at the same level as personality or social systems. Cultural systems are constituted by the organization of values, norms and symbols which guide decisions made by actors and, at the same time, restrict the types of interaction occurring between actors. Thus cultural systems are not empirical systems comparable to personality or social systems, but systems of cultural patterns at a higher level of abstraction whose different linked elements constitute value and interpretation systems

---

[7]In his early theoretical draft (The Structure of Social Action) Parsons assumed a dual differentiation of action systems (orientation to a rational purpose and categorical-normative orientations). This scheme is now developed into a triple concept. Later he condenses his action concept into the so-called AGIL scheme. If one follows Luhmann's interpretation, one can see in it a further development without contradiction (Luhmann 1984). The question of whether insisting on human subjects as actors in action systems is proof of animism which is different from Parson's approach (Luhmann 1988) does not interest us at this point. What appears important to us is the certainty that we have not removed a theoretical mosaic piece which is atypical for the entire structure and perhaps revised in later work. Luhmann confirms that this is not the case (Luhmann 1988; see also Münch 1988).

as well as systems of expressive symbols (Parsons, Shils 1962, 54f). For action system theory personality and social systems as well as culture are three integrated (sub)systems which belong together and can only be separated from one another analytically. In the analytical center of personality and social systems actors are presented whose actions/interactions can be regarded as purposefully, situationally and normatively regulated. The analytical approach to action systems from the cultural perspective is particularly significant because "systems of value standards (criteria of selection) and other patterns of culture, when institutionalized in social systems and internalized in personality systems, guide the actor with respect to both the orientation to ends and the normative regulation of means and of expressive activities, whenever the need-dispositions of the actor allow choices in these matters" (Parsons, Shils 1962, 56).

The binding nature of the value standards and cultural norms for action systems is created through institutionalization processes in the form of a generalized symbolically expressive value system. The latter guarantees the persistence of a basic pattern of binding cultural values while at the same time ensuring variation possibilities. The variability is to be considered as a condition for action autonomy. It comes about by virtue of interpenetration, whose typical form for Parsons is represented by the internalization of normative cultural elements or institutionalized cultural patterns in the personality as conveyed via socialization processes. In summarizing Weber's and Parsons' approaches, a classification and functional scheme for innovation and production action can be worked up which might have the following form (see Figure 1).

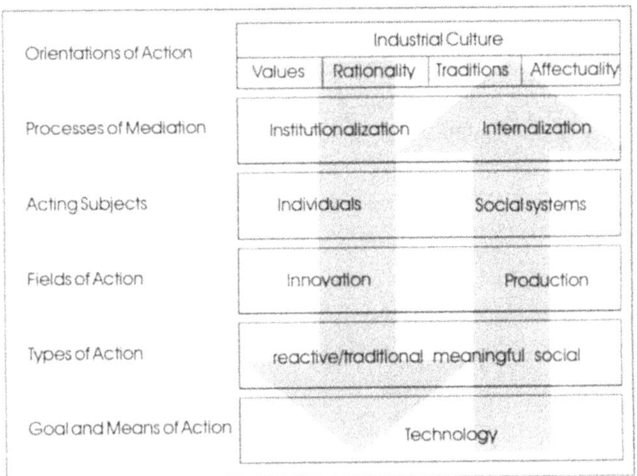

Figure 1: Scheme for innovation and production

The outlined scheme can be read in two directions, which are indicated by the arrows. If one reads the scheme from top to bottom, then steps of transmission emerge which start from culture or industrial culture, influence actions via various levels of authority and processes and in this way can also shape technology. This would be an action-constituting process, which up to now was the main research interest of the research network CAPIRN. The interpretation pattern, however, also includes the contrary form of interaction, i.e. the influencing of culture through technology. Technology, the artifacts as well as the processes of their creation and application influence culture/industrial culture. However, innovation and production are not the sole and possibly not the primary fields of action for this point of view (the field of everyday techniques is probably of greater significance here). Nevertheless, both lines of interpretation must be included in a truly dialectical approach in culture and technology, because the notion of two independent functions is just an analytical differentiation. In reality the relationship between culture and technology is an interactive and historical one. At all levels (of the above figure), both, culture and technology act upon each other and both lines of mediation are always simultaniously effective-in differing historically varying compositions.

So much for an initial overview of the connection between industrial culture and technology. In summary, it must first be pointed out that a deeper insight into transmission mechanisms, which can be applied to the question of industrial culture, was able to be gained through the reconstruction of classical concepts of action and culture. Nonetheless, the limitations of this approach must not be overlooked. A major difficulty lies in the concept of action, which both Parsons and Weber conceive as non-emphatic. This is not intended to be a reproach, but an indication of the area in which I propose to expand the theory. For the technology shaping approach which I favor, an emphatic, emancipatory concept of action which includes, on the one hand, production action organized and practiced according to anthropocentric aspects and, on the other hand, human-centered shaping of technology is essential. It is obvious that further theoretical development is necessary at this point.[8] On the one hand the classical connotation of action must be supplemented by elements of communicative action la Habermas (Habermas 1981); closely linked with the problematic concept of action is the unclear definition of subjectivity in action systems.[9] On the other hand it seems to be inevitable to enlarge the network of orientations by interest-guided action motives, because interests are not fully covered by a rationality of calculated purposes (in German language: Zweckrationalität).[10]

In the following paragraph I move on to a definition of the content-related dimensions of the concept of industrial culture.

---

[8]An outline of the problems and initial steps towards providing concrete contents can be found in the Synthesis Report on the Prospects of Anthropocentric Production Systems: A World Comparison of Production Models by Rauner/Ruth, 1991.

[9]Thus Luhmann (1988), for example, refers to Parson when he characterizes action theories which claim that human actors are a "subjective" requirement for action systems as animistically inappropriate versions of the action system theory; at the same time Parsons' approach in the form of the 'General Theory of Action (TGTA) can justifiably be interpreted as cybernetically subjectless just as well as subjectively "animistic".

[10]. Because of the limited space of this articla, the fields of further theoretical development can only be designated here.

5 A lot of dimensions ... in search of a consistent theoretical structure

With the aim of studying various technological developments, in an initial approach to the area under study we developed the concept of industrial culture, which comprised a set of cultural, social and psychological factors in addition to economic, political and institutional variables. The most important criteria in the early project phases were plausibility and capability of operationalization for empirical research. First of all, we classified the area under study. The following structured diagram provides an overview of the relevant dimensions and fields of analysis.[11]

a. Social Institutions

Here we include a basic stock of common cultural values and attitudes, such as the attitude of the population, particularly that of the workers concerned, toward technology and toward work or labour. Among the important social institutions are industrial relations with the two extremes, conflicting and cooperative relations, as well as the specific form of the social constitution of companies, i.e. a network of norms, rules and routines that influence the social system within the company in addition to the formal structures. Two institutions are of importance for the sphere of technological development: (1) styles and traditions of technology tied up with innovation objectives (e. g. technological leadership) depict the above mentioned social attitude toward technology in the materialized artifacts. And, in this connection views restricted by occupational socialization, which leave their mark as problem-solving perspectives of engineers (2), also emerge.

b. Industrial Organization

The industrial organization refers to the state, structure and organization of industry. Sector-specific, company-specific as well as department-specific elements of the control structures fall into this category as do national, sector-related or typical company organizational preferences, job classification systems and task allocation structures. The above mentioned structural features can be classified into the dichotomous poles, Taylorist/fragmented and work-oriented/decentralized-holistic, and thus furnish an important dimension for a content-related determination of the quality of industrial cultures. The industrial organization has special significance for the sphere of application of technology.

c. Educational Institutions

[11] The following sketch is a revised short version of the structural dimensions of industrial culture depicted in the article of Rauner/Ruth in this publication.

We consider the form and structure (of national) educational institutions to be an essential, shaping factor for industrial cultures. The relationship between the individual components of the educational system, their interaction and social foundation, such as in the form of a state-supported labour-force-requirements-approach or as the antithesis in the form of an individualized human capital approach, give this key dimension of industrial culture its specific character. The patterns of skill formation processes as well as the systems of vocational education/training are of central importance for the processes of design and use of technology.

d. Research, Technological and Structural Policy

State action in the political fields of research and technology can be oriented in very different ways regarding its content, with (financing) focus on basic technologies and large-scale projects, for example, or oriented to the needs of small and medium-size enterprises. In the area of promotion of production technology the emphasis can be placed on very different aspects (technology-oriented or work-oriented), which contribute to a specific industrial culture and thus have far-reaching impact on the design and use of technologies.

e. Psychology

The psychological influencing variables include socializational and subject-constituting factors, and aspects of social as well as individual psychology. Culturally varying types of socialization find expression to an especially large degree in the field of vocational socialization, which is important for technology-related action processes. In addition, the various cultural and social patterns of constituting subjectivity just as the significance and kind of relationship between individual and society are of importance for technological development and application processes to be interpreted for industrial culture. A factor influencing industrial culture, especially technological development, which should not be neglected is that of the dreams of technology-shaping engineers. Anticipatory daydreams, pipe dreams and - now and then one has the impression that this type of dreaming must have preceded many technical artifacts - nightmares differ very greatly from each other in various cultural contexts. The dreams mix with images of technology or themselves contribute to the constitution of various industrial culture images of technology.

As a result of this classificational structuring of industrial culture into five dimensions or fields of analysis, the most varied cultural, social and other (-operationalizable) factors are combined. This complex is, in turn-and that is what is actually significant about the approach-based on industrial culture values which I construe as bipolar pairs of opposites. The contrary concepts are in a framework, within which the individual industrial culture variables or industrial cultures can be classified on a "scale". The following value pairs are regarded as relevant for the present status of the approach:

social integration - social segregation
individualism - groupism

orientation to rules - orientation to principles
analytical/segmenting - synthesizing/holistic

Admittedly, these pairs of opposites are tailored to the contrast between oriental and occidental culture. This may be criticized, but the discussions revolving around production concepts and industrial performance seem to boil down to this very contrast.

The cultural value pattern was essentially obtained deductively, i.e. it is based on relevant approaches of comparative studies. The pair of opposites, individualism - groupism, is based on empirical research in the context of economic culture, specifically on the work of Bond and Hofstede (1990) regarding the connection between culture and economic performance. The term groupism must not be misinterpreted in the idealized Western sense of group work which stands for human-centred work organization in which the individual worker can acquire self-confidence and become emancipated. In the Asian and particularly in the Japanese case groupism and work groups mean to be part of the whole. Indeed, Japanese work groups seem to have little or no communication barriers; a fact which is so fascinating for Western social scientists. This communicative climate is not in alignment with individualism and self-realization but more likely it is adjusted towards the subordination under the group's (or company's) weal and woe.

The pair of opposites, orientation to rules vs. orientation to principles, is based on the results of a comparison between German and Japanese styles of design. This study makes use of differing philosophical and cultural traditions to explain differences in design styles and the development of technology. (Technical) problem-solving processes oriented more to theory and principles are derived from a cognitive orientation to the understanding and derivation of knowledge from principles in Germany (essentially based on German idealism). In the Japanese version Confucian, greatly rule-oriented thinking structures find expression in the design processes as algorithmizing work flow plans with emphasis on experiments and systematic trial and error processes for obtaining optimum solutions (Moritz 1991). In a similar fashion the concept pairs, social integration-social segregation as well as analytical/segmenting- synthesizing/holistic, are based on philosophical traditions.[12]

The consequences of this basic cultural orientation of values can be worked out, for example, on the basis of how they manifest themselves in educational institutions. The human capital model of educational institutions corresponds more to an industrial culture having a strongly individualistic character while a model of a socially guided labour force requirements approach corresponds to an industrial culture greatly characterized by groupism. Without question the U.S. can be regarded as a prototype of the first model, i.e. of a basically individualistic character and thus in accordance with a personalized human capital investment model. Whereas the model can be found everywhere in the U.S. from general to vocational education, the exemplary allocation of a groupist industrial culture to the antithesis, the labour force requirements

[12]The development of this connection must be left out for reasons of space. I refer to forthcoming articles.

approach, is more difficult.

There is hardly any doubt about the dominance of groupist values in Japan's industrial culture, though a clear allocation to the labour force requirements approach cannot be made. Rather, a differentiation into general and vocational education results in a greater orientation to the labour force requirements approach in the case of the latter while the human capital model seems to be more appropriate for general education. It appears even more difficult in the case of German industrial culture. Compared to American and Japanese industrial culture, the latter is located closer to the individualistic pole on the scale, but the dominant educational institution must be regarded as more groupist and less individual-oriented. In these examples one can clearly see the difficulties related to the model of pairs of opposites, i.e. that the polarities do not apply in their pure form, but overlap and influence each other. It also becomes evident how difficult it will be to provide empirical verification of the connections between bipolar cultural orientations and the specific character of industrial culture variables. Nevertheless, there is a lot to be said for this model. If one takes into account the reciprocal influence of the industrial culture dimensions and the resulting lack of clarity in the determination of interrelations, then the interaction between industrial culture, industrial culture dimensions and variables as well as fields of action can be described as follows. The connections are taken together in an analytical perspective and can be interpreted in a converse line of argumentation as a constitution process.

Industrial culture must be comprehended as a framework, as one (Parsons-like) cultural system (in addition to other conceivable ones, such as everyday culture), which provides fundamental values and orientations in the form of cultural value patterns. It can be divided into five fields of analysis or dimensions, to which empirical research is oriented. Every field of analysis is composed of a set of variables, via which industrial culture influences action systems for the development and application of technology (see Figure 2).

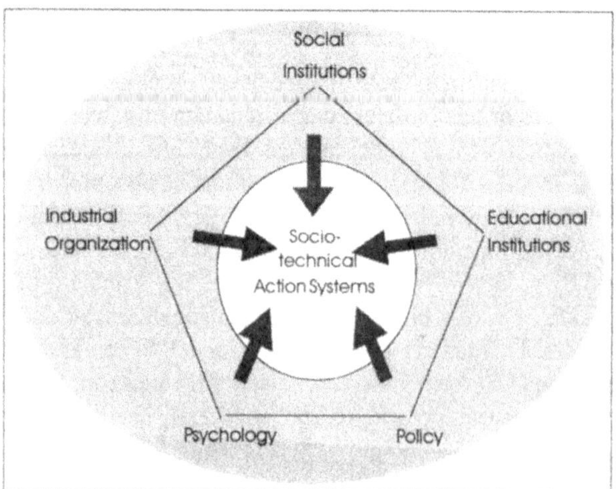

Figure 2: Industrial Culture and Socio-technical Action Systems

The concept of real technology, with which all actions for the creation and application of technology in addition to technical artifacts are comprehended, is used as the basis for action systems. Thus real technology is directly related to action systems which are then interpreted as socio-technical action systems, i.e. systems in which people deal with technical artifacts and technical systems (Ropohl 1979 and 1988). The question then arises as to the mechanisms by means of which acting people adopt industrial culture values and how the latter are transferred to action orientations. I assume institutionalization processes, i.e. the coagulation of cultural contents into social structures, and internalization, e.g. the socializational sedimentation of culture and social contents in personality systems and finally in actions, to be effective steps of transmission between industrial culture and socio-technical action systems.

6 Conclusion and Outlook

In conclusion, I want to take a last look at the cognitive interests and application connections of the industrial culture approach developed here. The special feature of the approach is the fact that it is a comparative concept, i.e. industrial cultures, industrial culture variables and their special effects on actions systems are not fully developed until they have been classified and evaluated in a comparative analysis going beyond the plausibility and relevance inherent in the respective industrial culture. This, of course, thus rules out the singular industrial culture in the sense of a worldwide, standardized (or standardizing) industrial culture which might then be conceived as the superior world production and innovation model. Just the opposite is true: the fundamental axiom of the industrial culture approach assumes the existence of many industrial cultures, each of which has its own strengths and weaknesses as well as its own dynamics of development, without having to come closer together in a worldwide convergence. The approach does adequately take into account converging tendencies, but not in the way in which modernization theories conceive this as an inevitable, extensive alignment of industrial cultures with each other.[13]

The epistemological objectives of empirical industrial culture research are pursued in two ways: via a micro- and via a macro-perspective. The micro-orientation looks at questions of the influence of industrial culture on the development and application of technology. This line of research includes the questions as to the connection between industrial culture variables, such as skilled work, and the development of specific shop floor oriented CNC controls; or a historical example: the connection between engineering traditions and the content-related orientation of engineering training as well as special structural features of steam engine production in different countries (see Radkau 1989). Finally this line of research compares the specifically formed relations between industrial culture values with technology, its development

[13]Details on convergence/divergence cycles can be found in Rauner, Ruth 1991.

and application in various industrial cultures.

In the macro-perspective research is focused on industrial culture overlapping processes (acculturation). From this point of view great importance is attached to the problem of the application or applicability of industrial cultures or parts of them. This is also linked to the question of the different possibilities open to individual industrial cultures in coping with creative adaptations of singular industrial culture elements. The conditions for successful intercultural learning processes as well as their special forms and contents are the central subjects of research from a macro point of view of the industrial culture approach.

By virtue of the last mentioned aspects, a link is forged to the current discussion regarding lean production. The industrial culture approach might offer a more solid scientific basis for the, up to now, very intense discussions on lean production in general and on the question of applicability in particular.

References

Berger, P. L. (1987). The Capitalist Revolution. London.

Bond, M. H. and G. Hofstede (1990). The Cash Value of Confucian Values. In: Clegg, St. R. and S. G. Redding (Eds.) ass. by M. Cartner, Capitalism in Contrasting Cultures, p. 382-390. Berlin, New York.

Clegg, St. R., Higgins, W. and T. Spybey (1990). ' Post-Confucianism', Social Democracy and Economic Culture. In: Clegg, St. R. and S. G. Redding (Eds.) ass. by M. Cartner, Capitalism in Contrasting Cultures, p. 31-78. Berlin, New York.

Clegg, St. R. and S. G. Redding (1990). Introduction: Capitalism in Contrasting Cultures. In: Clegg, St. R. and S. G. Redding (Eds.) ass. by M. Cartner, Capitalism in Contrasting Cultures, p. 1-28. Berlin, New York.

Crow, M. C. and B. Bozeman (1991). The National System of Innovation in the United States. Paper prepared for the 2nd Conference on Policies and Strategies on Technology in Industrialized Countries in Moscow, May 1991.

Dosi, G., C. Freeman, R. Nelson, G. Silverberg and L. Soete (Eds.) (1988). Technical Change and Economic Theory. London.

Dosi, G. and L. Orsenigo (1988). Coordination and Transformation: an overview of structures, behaviours and change in evolutionary environments. In: Dosi, G. et al (Eds.), Technical Change and Economic Theory, p. 13-37. London.

Freeman Ch. (1987). Technology Policy and Economic Performance. London.

Gottl-Ottlilienfeld, F. von (1914). Wirtschaft und Technik. In: Grundriss der

Sozialökonomik, II. Abteilung. Tübingen.

Hirsch-Kreinsen, H. (1989). Entwicklung einer Basistechnik - NC-Steuerungen von Werkzeugmaschinen in den USA und der Bundesrepublik. In: Düll, K. and B. Lutz (Eds.), Technikentwicklung und Arbeitsteilung im Internationalen Vergleich, p. 161-211. Frankfurt/München.

Hofstede, G. (1980). Culture's consequences. London.

Hughes, T. P. (1983). Networks of Power. Baltimore/London.

Kuby, Th. (1980). Vom Handwerksinstrument zum Maschinensystem. In: Arbeitspapiere der TU Berlin. Berlin.

Layton, Edwin T. (1978). Millwrights and Engineers, Science, Social Roles, and the Evolution of the Turbine in America. In: Krohn, W., F. T. Layton and P. Weingart (Eds.), The Dynamics of Science and Technology. Sociology of the Sciences, Vol. II, p. 61-87. Dordrecht.

Luhmann, N. (1984). Soziale Systeme. Frankfurt/Main.

Luhmann, N. (1988). Warum AGIL? In: Kölner Zeitschrift für Soziologie und Sozialpsychologie Vol. 40, p. 127-139.

Lundvall, B. A. (1988). Innovation as an interactive process: from user-producer interaction to the national system of innovation. In: Dosi, G. et al (Eds.), Technical Change and Economic Theory, p. 349-369. London.

MIT Commission on Industrial Productivity (1989). The US Machine Tool Industry and its Foreign Competitors. In: Working Papers of the MIT Commission on Industrial Productivity, Vol. II. Cambridge, Mass.

Moritz, E. F. (1991). Konfuzius - Japan - Technik. Ein alter Hut neu aufgesetzt. In: Jahrbuch des Deutschen Museums 1991. München.

Münch, R. (1988). Theorie des Handelns. Frankfurt/Main

Mumford, L. (1934). Technics and Civilization. New York.

Murikami, Y. (1986). Technology in transition: Two Perspectives on Industrial Policy. In: Patrick, H. (Ed.), Japan's high technology industries: Lessons and limitations of Industrial Policy, p. 211-241. Seattle, London.

Noble, D. F. (1984). Forces of Production. A Social History of Industrial Automation, New York.

Parsons. T. (1968). The Structure of Social Action. New York (1st Edition: 1938).

Parsons. T. et al. (1962). The General Theory of Action. In: Parsons, T. and E. A. Shils (Ed.), Toward a General Theory of Action, p. 3-29. Cambridge, Mass.

Parsons, T. and E. A. Shils with assistence of J. Olds (1962). Values, Motives, and Systems of Action. In: Parsons, T. and E. A. Shils (Ed.), Toward a General Theory of Action, p. 47-275. Cambridge, Mass.

Radkau, J. (1989). Technik in Deutschland. Vom 18. Jahrhundert bis zur Gegenwart. Frankfurt/Main.

Rammert, W. (1988). Technisierung im Alltag. Theoriestücke für eine spezielle soziologische Perspektive. In: Joerges, B. (Ed.). Technik im Alltag, p. 165-197. Frankfurt/Main.

Rammert, W. (1988a). Technikgenese. In: Kölner Zeitschrift für Soziologie und Sozialpsychologie, 40, 1988, p. 747-761.

Rauner, F. and K. Ruth (1989). Industrial Cultural Determinants for Technological Developments: Skill Transfer or Power Transfer? In: AI & Society, Vol 3, No. 2 p. 88-102. London.

Rauner, F. and K. Ruth (1990). Perspectives of Research in 'Industrial Culture'. In: Karwowski, W. and M. Rahimi (Eds.), Ergonomics of Hybrid Automated Systems II, Amsterdam.

Rauner, F. and K. Ruth (1991). The Prospects of Anthropocentric Production Systems: A World Comparison of Production Models. Brussels.

Ropohl, G. (1979). Eine Systemtheorie der Technik. Zur Grundlegung der Allgemeinen Technologie. München, Wien.

Ropohl, G. (1988). Zum gesellschaftstheoretischen Verständnis soziotechnischen Handelns im privaten Bereich. In: Joerges, B. (Ed.). Technik im Alltag, p. 120-144. Frankfurt/Main.

Rosenberg, N. (1982). Inside the black box. Technology and economics. Cambridge, Mass.

Schelsky, H. (1965). Der Mensch in der wissenschaftlichen Zivilisation. In: Schelsky, H. (Ed.), Auf der Suche nach Wirklichkeit, p. 439-480. Düsseldorf, Köln.

Ullrich, O. (1979). Technik und Herrschaft. Frankfurt/Main.

von Hippel, E. (1988). The Sources of Innovation. New York.

Weber, M. (1972). Wirtschaft und Gesellschaft. Tübingen (5th revised Edition, edited by J. Winckelmann; 1st Edition: 1922).

Wilks, St. (1990). The Embodiment of Industrial Culture in Bureaucracy and Management. In: Clegg, St. R. and S. G. Redding (Eds.) ass. by M. Cartner, Capitalism in Contrasting Cultures, p. 131-152. Berlin, New York.

Section II

Empirical Investigations of Competitiveness of the Machine Tool Industry in Different Industrial Cultures

Research and Development Activities to Enhance Market Competitiveness of Products in Japanese Machine Tool Industry

by Yoshimi Ito

1 Introduction

The most characteristic feature of the present production environment can be represented by the combination of the following three major key terms, i.e., globalisation, international co-operation and international competitiveness. In this regard, there is an argument on what are the desirable products and production facilities for the already industrialised and industrialising nations? In accordance with some reports, which have described the predicted results for the future production environment[1], it is very easy to enumerate such products, e.g., strategic product, highly value-added product, human sensitivity-oriented product and region-specified product[2]. When referring to these products, people can have an idea that the marketability of the product becomes very important.

This marketability can be interpreted as to be the market competitiveness of the product when considering the fierce competition in the world market, even though there is a beautiful term apparently describing the importance and inevitability of the international co-operation.

One questioned point is, thus, how to enhance the market competitiveness of the product through the research of the related technology and also the product development within a nation.

To answer this question, it is necessary to carry out a considerable number of the literature surveys, questionnaire investigations, on-the-spot investigations and so on along with the insightful analyses. This is because the related activities cover very wide area ranging from the market survey to the secondary level-improvement of the product, and also from those of governmental project research to small-size co-operation between the university and the enterprise. In addition, it is very difficult to assess the effects of the indirectly functioned activities, e.g., completely academic-based works.

In this paper, author tries to explicate the present trends of the research and development activities in the Japanese machine tool industry and related organisations, which are directly and indirectly concerned with the enhancement of the market competitiveness. Concretely saying, the paper describes the organisational structure, finances, research subjects and characteristic features of each activity.

2 Overall View of Research and Development Activities

The production technology has amazingly evolved, especially in the field of flexible manufacturing along with upheaving some new waves, and these new waves can be represented by the following key terms. For instance, "Production of Sustainable Growth", "Production Structure with International Co-operation", "Lean and Compact Manufacturing", "Ecology-Oriented Production" and "Thought ModelBased Manufacturing (THOMAS)[2]". As literally shown, these new waves aim mainly at the establishment of the production tool, which could generate the highly value-added products under the internationally co-operative environment; however, most of people interpret it as follows.

"The production technology is still the most powerful weapon to cope with the fierce competition in the international market, simultaneously expecting to produce the wealth for their society".

This opinion is in part credible, and as a result the machine tool industry has been, in turn, required of enhancing the marketability of the product, because the machine tool is considered as to be a crux of the manufacturing facilities.

In brief, the present machine tool technology should cover very wide area, ranging, for instance, from the analysis of the deep knowledge of mature engineer[3] to the development of the transducer for the in-process measurement. In due course, an incredibly large number of the activities on the machine tool and related technologies is being carried out by various organisations. Generally, these activities appear not to have any correlations, but to carry out independently by each organisation. This understanding is completely wrong. In fact, the Japanese research and development activities are activated by the well-configured human network, which is, in principle, organised on the basis of "Alma Mater" and also of the close society called "Machine Tool Engineer's Family". In other words, the high qualification of Japanese machine tool technology lies on the strong human network to a large extent, which is centred by the academic staffs of several leading universities and related organisations. Although being not explicit, this fact is one of the marked points and derived from the historical background in the educational and training systems of the machine tool engineer. In addition, the Japanese machine tool society is small, and also has an implicit system to transfer the grass root-like knowledge, information and technology from the senior to junior engineers as like as the "Inheritance from Father to Son" without using any documentations. This evidence is difficult to understand correctly especially for the foreign people; however, such knowledge are one of the basic prerequisites when investigating the reasons of why the Japanese machine tool is so powerful within the world market.

Figure 1 demonstrates duly such a map illustrating the overall view of the research and development activities in the Japanese machine tool society after making some simplifications. As shown in this figure, the Japanese society conducts vigorously the research and development activities from various aspects, and a part of these activities is functioned directly to enhance the competitiveness of the product. In this context, there are the two types of the activities. One is that for supporting or backing up the product development and another aims at the direct enhancement of the market competitiveness of the product. Generally, the former is carried out in the form of the co-operation among the governmental departments including the national research laboratories, semi-national body organisations, professional societies, universities and enterprises. In contrast, the latter is mainly carried out by each company itself.

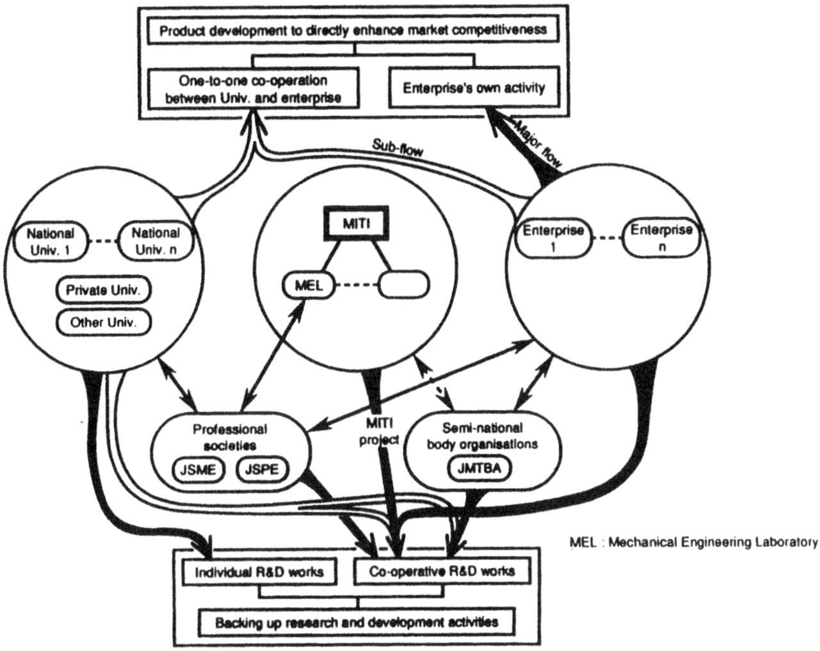

Fig. 1 A Map Illustrating Overall View of Japanese Research and Development Activities

3 Backing up Research Works and Organisations to provide Machine Tool Manufacturer with highly qualified Capabilities for Product Development

Apart from the activities directly conducted by each machine tool manufacturer itself, almost all the research and development works can be, as stated in the preceding Chapter, regarded as to be the supporting ones. Generally, the aims and scope of the supporting work can be characterised by its major organiser, and every supporting work has its own organisation, financial size, objective subjects and expectable final out-put. In the following, details of some supporting works will be described along with the typical case studies, because it is, of course, very difficult to point out all the supporting works without exception.

3.1 MITI Projects

The MITI (Ministry of International Trade and Industry) has often promoted the cultivation and fertilization of the machine tool industry by launching out some projects, e.g., Big National Project, Small-and Medium-Size Project, Internationally Cooperative Research Project and so on. Generally, these projects cover over very wide technological areas and sometimes involve the machine tool and related technologies as a part of it. Meanwhile, there have been some rumors that the MITI controls strongly the strategy, scheme, research plan and schedule of the project from political point of view. This rumor is not true, but the MITI endeavours to prepare the research and development environment by, for instance, supporting a part of research fund after making the tough negotiation with the Ministry of Finances.

In principle, the MITI projects have commonly the following marked features.

(1) The project is performed by the semi-national body organisation, which consists of the staffs and members came from the enterprise, national research laboratory and university.

(2) The company member manages the activities, whereas the university member plays the role of intermediator and advisor.

(3) The research and development work is planned to reach about 50% of the completeness of its final goal, i.e., the common or basic portion of the related technology being objective.

For instance, Fig. 2 reproduces the basic layout of the large-size FMS to produce the axial plunger pump of swash plate type[4].

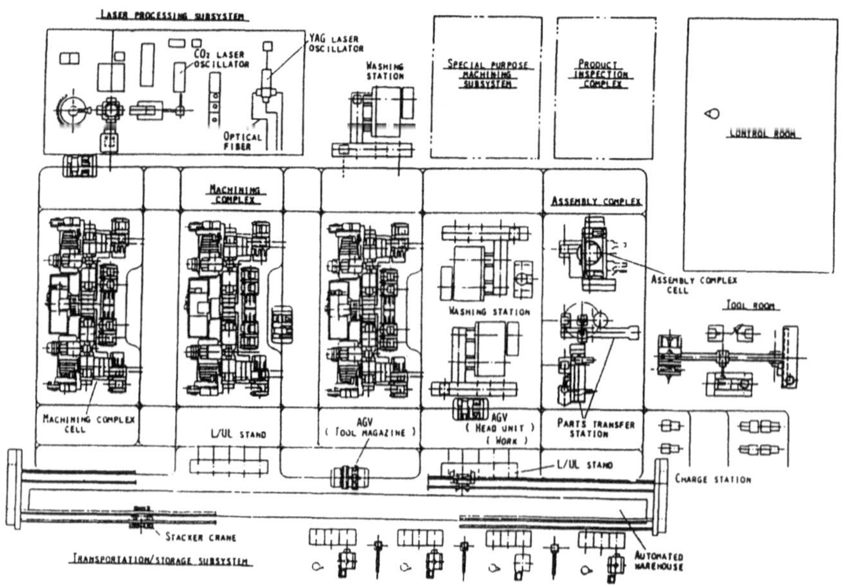

Fig. 2 Basic Layout of Large-size FMS to Produce Axial Plunger Pump of Swash Plate Type

This layout was produced by the Engineering Union of Big National Project named "Flexible Manufacturing System Complex Provided with Laser (FMSC)". Based on it, a company has later made the detailed design and installed a complete system as manufacturing facilities. To have some ideas, Table 1 summarises, furthermore, the outlines of this project.

<table>
<tr><td colspan="2">Research period</td><td>8 years (1977-1985)</td><td rowspan="6">Core research subjects</td><td rowspan="6">Design, manufacture and operation of test plant

Machining technology of multiple-function type

Flexibly advanced raw material processing

Flexible assembly system

Laser application technology

Automatic diagnosis systems

CAD / CAPP / Process control technology</td></tr>
<tr><td colspan="2">Research fund</td><td>≒13.7 Billion Yen</td></tr>
<tr><td colspan="2">Aims and scope</td><td>To develop the complex system for batch production positively using laser technology</td></tr>
<tr><td rowspan="3">Research / development organisation (Engineering union)</td><td>Companies</td><td>20</td></tr>
<tr><td>National research laboratories</td><td>3</td></tr>
<tr><td>Members from universities</td><td>11</td></tr>
</table>

Table 1 Outlines of Big National Project "FMSC"

Generally, the MITI project is functioned to catch up the leading nation's technology, and also to step our advanced technology out to the new stage. For instance, the Project "FMSC" has been evaluated as follows.

"Japanese flexible manufacturing technology was advanced amazingly by the FMSC project, resulting in the provision of the sophisticated manufacturing facilities and the enhancement of the marketability of the FMS (Flexible Manufacturing

System). In addition, there were considerable repercussion effects on the academic research activities for the FMS and CIM (Computer Integrated Manufacturing). For instance, the prediction activities on the future production environment are very excellent and superior to those of other countries, showing the upmost high qualification in the proposals of the future system concept and due research subjects[1, 5]".

In due course, this superiority in the future prediction could reflect to the enhancement of the marketability of the FMS and CIM hereafter by accomplishing the leading-edge research and development works.

3.2 JMTBA Activities

The JMTBA (Japan Machine Tool Builders' Association) has conducted various activities so far, for instance, "JIMTOF (Japan International Machine Tool Fair)", "IMEC (International Machine Tool Engineer's Conference)", "International Seminar for Technical Information Exchange", "Standardisation Activities of JIS and ISO", "Application-Aided Research and Prototype Development Project" and so on.

These activities are, in general, carried out in co-operation with the member companies, national laboratories and universities, mainly aim at the dissemination of the corresponding information, and in part intend to enhance the international co-operation.

Of these, dare to say, the application-aided research and prototype development project has some flavour of reinforcing the international competitiveness. This project is generally carried out by forming the task force mainly consisting the company members, although the conflict among companies is very severe in Japan. Table 2 summarises the title of subject, chairperson and research period of each project so far performed.

Period (Year)	Subject	Chief of task force
1981～1983	New structural material (Application of concrete to structural body)	Prof. Y.Ito (Tokyo Institute of Technology)
1983～1984	Pattern recognition technology	Prof. T.Honda (Tokyo Univ.of Agriculture & Technology)
1984～1985	Development of new flexible chuck — Top jaw quick changing mechanism	Prof. Y.Ito (Tokyo Institute of Technology)
1985～1987	Thermal deformation and its remedies	Prof. Y.Yoshida (Chiba University)
1988～1991	Basic design technology (Development of machine tool with super-multiple function / Enhancement of reliability using new materials)	Prof. Y.Yoshida (Chiba University)
1991～	Structural design technology along with formulation of design data	Vice director Y.Tatsue (MEL of MITI)

Table 2 Application-aided Research and Prototype Development Projects of JMTBA

Generally, these projects are carried out using the following philosophies.

(1) The objective subject is chosen in considerations of both the common interest of the member companies and the urgency of its development or improvement.

(2) In the case of the prototype development, the completeness rate of a machine or device is about 50% and the rest is under the responsibility of each company. In other words, the technology development is limited within its basically common portion.

(3) The chairperson is appointed from the academic staffs of the national laboratory or university so as to guarantee the fairness.

(4) The necessary finances are prepared using the seed money supporting system of the MITI and also JMTBA's own sources. In certain cases, the member companies are requested to last the special fund.

In the following, the outline of a project will be described to aid the deep understanding on what was ongoing in Japan. The 3 year-long project (1981-1983) entitled "New Structural Material" aimed at the development of the next generation MC (Machining Centre) by enhancing its thermal and dynamic performances, wherein the total expenditure was amounted about 24 Mil. yen. In this case, a prototype was produced by employing the leading-edge technologies at that time, although their practical applicabilities were in uncertainty[6]. Figs. 3, 4 and 5 show the technologies investigated, i.e., water cooling of ball screw, application of heat pipe to main spindle system and trial of stuffing the ferrite resin concrete in the structural body.

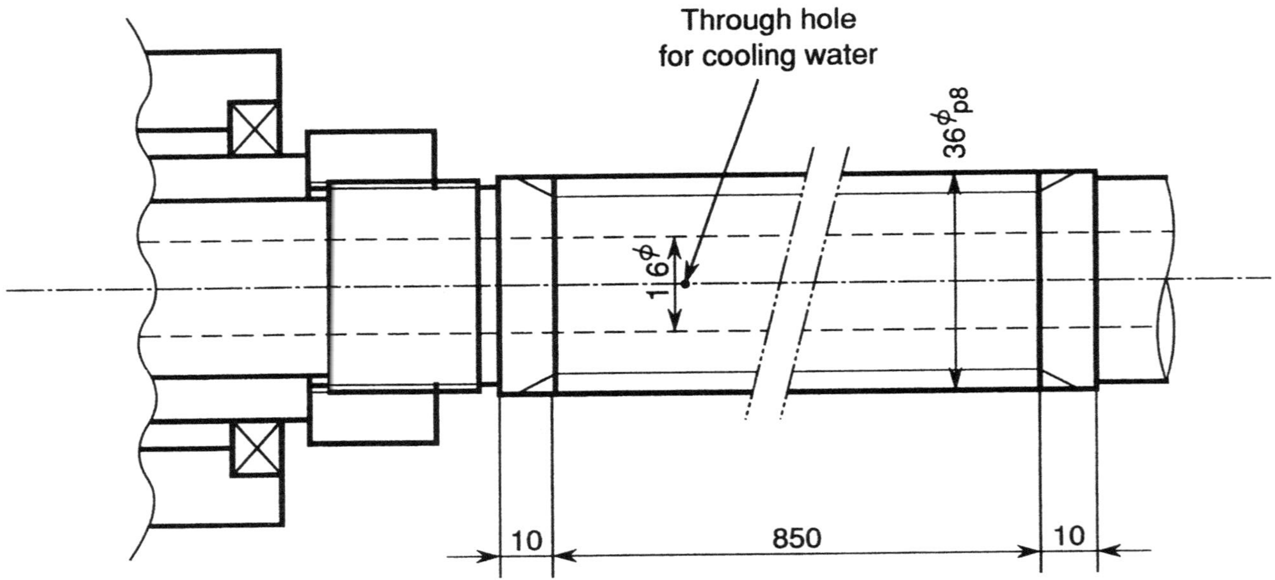

Fig. 3 Ball Screw with Water Cooling System

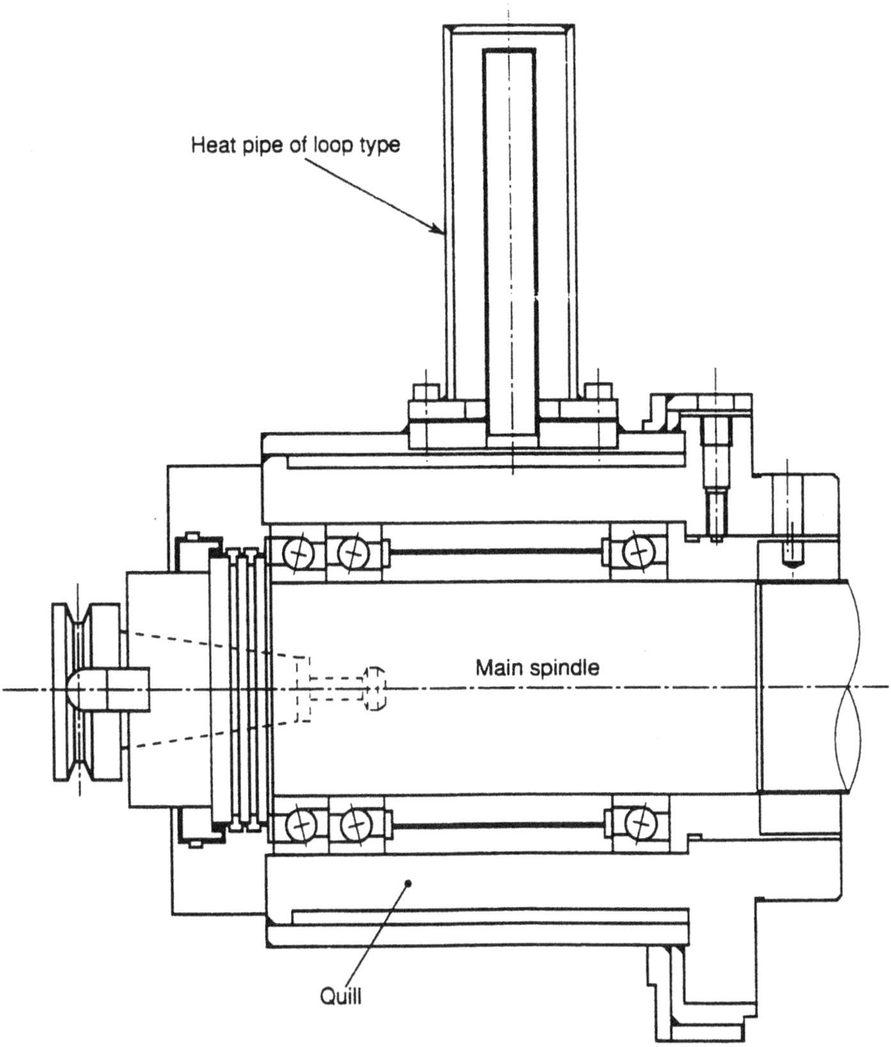

Fig. 4 Heat Pipe of Loop Type Applied to Main Spindle

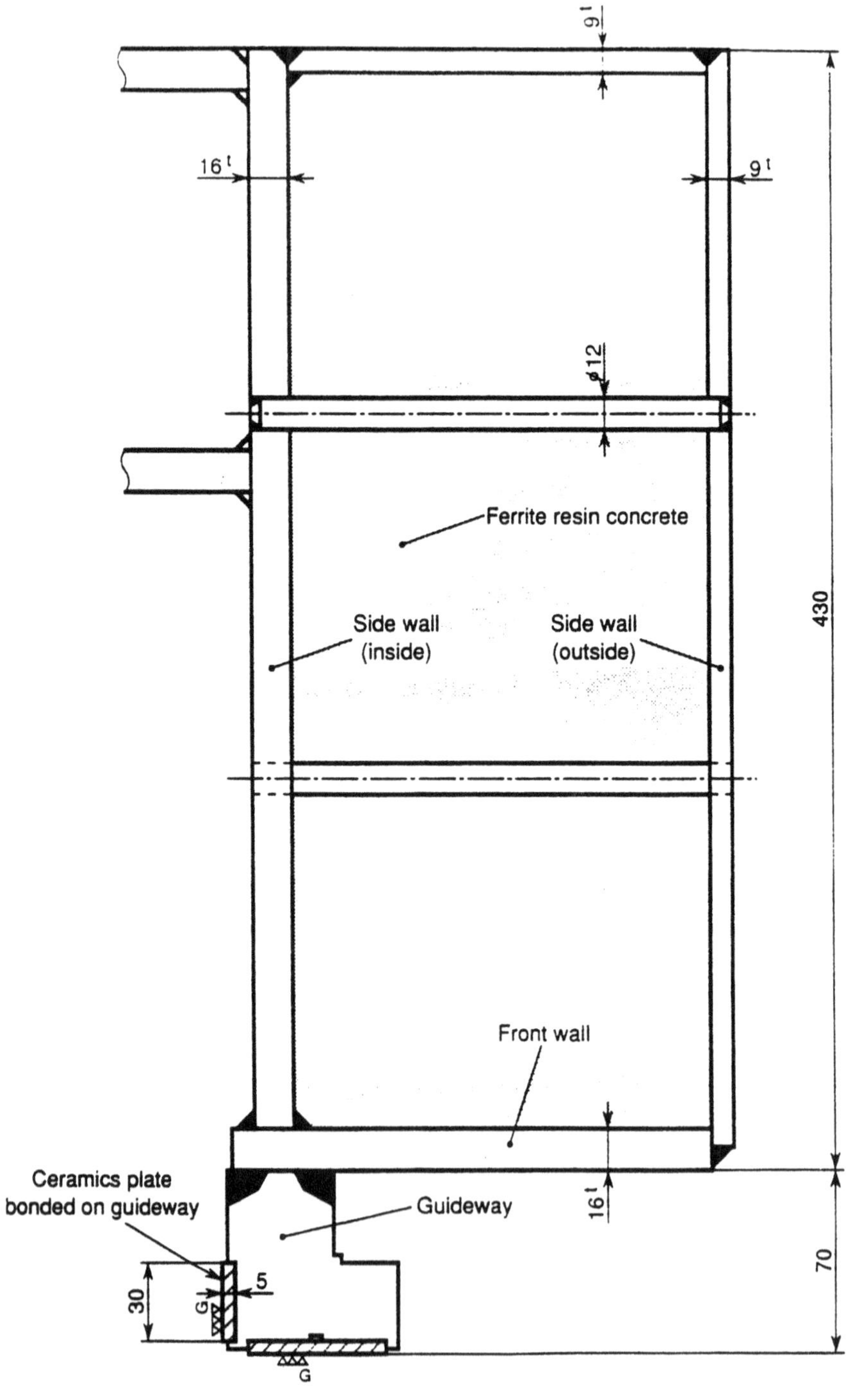

Fig. 5 Column Stuffed up by Ferrite Resin Concrete

In evaluation of the direct and indirect impacts of this project, now, the following impressions can be stated.

(1) The water cooling of the ball screw is now common and popular tool to reduce the thermal deformation of the driving system, resulting in the standard equipment of Japanese MC.

(2) The application of the heat pipe was not so effective to stabilise the thermal performance of the main spindle. It is, however, envisaged that Japanese machine tool manufacturer has been recognised the valuableness of the heat pipe. In addition, it is now possible, on the request of the user, to employ the heat pipe as one of the remedies to reduce the thermal deformation.

(3) Some of Japanese machine tool manufacturers are attracted to the application of the cement or epoxy resin concrete to the main body structure, and in fact, there is a certain number of machine tools made of such a concrete on the market. In contrast, the ferrite resin concrete is not employed as yet, because of its high raw material cost.

In addition, we must be aware of it that some projects were carried out by the devotional co-operation of a specified manufacturer, who produced a prototype without having any reimbursement. This is one of the reasons why the project can be performed with the expenses of relatively small amount.

3.3 Co-Operative Research Activities within Professional Societies

Apart from the regular activities on the meeting, publication and related works, the JSME (Japan Society of Mechanical Engineers) and JSPE (Japan Society of Precision Engineering) have some systems for the research unit to engineer the co-operative work among the university, national research laboratory and enterprise. As an example, Table 3 summarises the research units within the JSME, which have been carried out the co-operative works on the machine tool and related technologies.

Research unit	Major subjects	Research period (year)	Chairperson	Number of members	
				from research field	from company
Enhancement of stiffness and cutting performance of machine tools	Applicability of mechanical impedance to structural design Influence of vibration on machining accuracy Stiffness of machine tool joints Development of software for computing the stiffness	1970～1973	Prof. M. Honda	16	38
Automatic maintenance of machining system with DNC	Analysis of necessary functions in automatic maintenance system Development of in-process sensor with special reference to AC Performance analysis of machine tools	1973～1976	Prof. N. Takenaka	15	33
Functional enhancement of modular-designed machine tools and its evaluation method	Correlation between limiting depth of cut and stiffness Modular-designed structure Reliability and health diagnosis	1976～1979	Prof. H. Okushima	19	26
Man-machine system for clarifying machining environment	Ergonomics for machine tools Clean machining environment Increase of machining reliability with man-machine system	1979～1982	Prof. S. Yonezu	21	≒20
Machining system of next generation	Accuracy assurance and reliability New machine tool technology Interfacing technology	1982～1985	Prof. H. Takeyama	27	34
Leading-edge machining system	Extreme machining technology Ultra-precision machining Elements for machine tools	1985～1988	Prof. H. Sato	26	41
Fundamental technology for production enhancement	Manufacturing technology for new materials Ultra-precision technology of next generation Manufacturing technology of AI type	1988～1991	Prof. K. Iwata	25	34
Advanced machine tool systems and technology	Design technology for highly qualified structure Application of AI systems technology New machining technology / Multiple-function sensing	1991～1994	Prof. Y. Yoshida	25	35

(c)

Table 3 Co-operative Research Units so far Organised within JSME

As literally shown by the unit's name, the activities have covered over various aspects of the research subjects, and these are believed to fertilize each company's ability for the product development.

In this system, generally, the company member should pay the participation fee based on the regulation of the society, e.g., about 400 thousand Japanese Yen per year at present, whereas the non-profit members, i.e., university and laboratory members, can receive a certain amount of counterpart fund for lading the responsibility to conduct the research work. Some noteworthy points of this activity can be stated as follows.

(1) Non-profit members can choose their own research subject without having the strict limitation. In other words, the steering committee of the research unit settles only the perspective view of all the research activities, and then detail is under the control of each researcher.

(2) Company member can expect to receive the following advantages as a reimbursement.

(a) Collection of necessary information through the regular meeting and also the presentation of the research report by the non-profit member.

(b) Opportunity to have free consultation by the non-profit member.

Figure 6 shows the overall view of the research subjects being carried out in the present research unit.

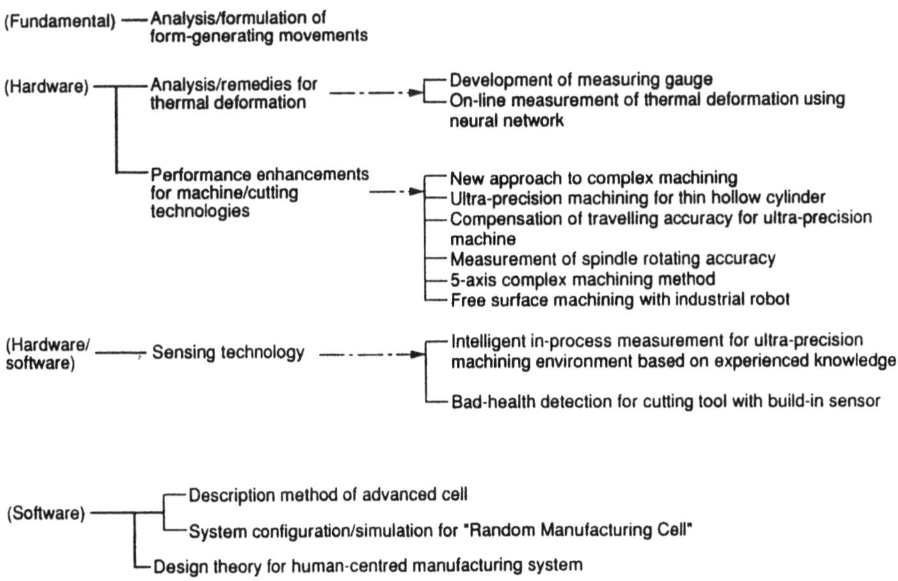

Fig. 6 Subjects Being Carried out in Co-operative Research Unit on Advanced Machine Tool Systems and Technology

3.4 Co-Operation between University and Company

There are the two types of the co-operation in this category. One is the co-operation based on one-to-one relation, and another is in the form of one university-to-plural companies. Generally, the former is organised so that the company can solve the facing problem using the technological and human resources of the university. Thus, this will be described in the succeeding Chapter.

For reasons of having a lot of well-configured professional societies and semi-national body organisations, the latter case is very rare in Japan. In this regard, an outstanding system is on work at the Tokyo Institute of Technology, i.e., "Creative Research Project of T.I.Tech.". Fig. 7 illustrates the organisational structure of this project, and as can be seen its cruces are as follows.

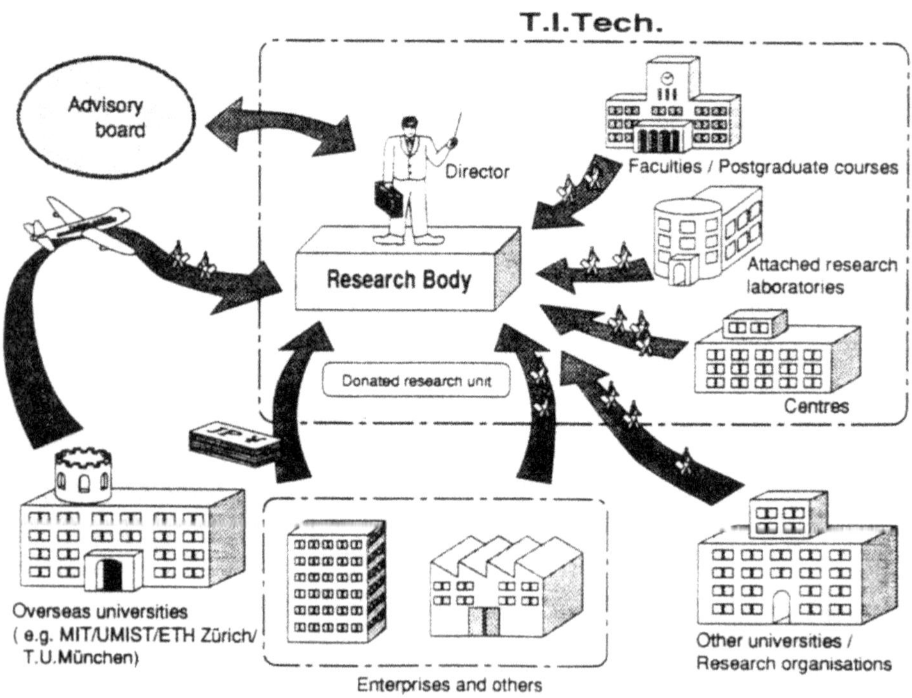

Fig. 7 Organisational Structure for Creative Research Project of T.I.Tech.

(1) The company provides the research fund and also, in certain cases, the man power.

(2) The Institute organises a project team, i.e., research body (research unit complex) by gathering the suitable staffs from the Institute itself, and, if necessary, those from other inland universities, research organisations, foreign universities and so on.

(3) The research subject is chosen in consideration of the long-range future trends of the objective field, if possible, those of about 10 years ahead. In addition, the subject is academic-oriented, not paying special attentions to the commercialisation.

At present, there are, as shown in Table 4, 7 projects and of these a 3 year-long

Research body (Pet name)	Research subject	Leader of body
	Next generation visual tele-communication with high quality	Professor S. Tsujii
	Development of photo-electronic beams device	Prof. K. Takahashi
	Healthy society systems of near future	Prof. H. Kukimoto
	Planning technology for infrastructure of highly integrated city	Prof. T. Kimura
	Science and technology compatible with greenhouse effect	Associate Prof. Y. Tamaura
CAD21	Computer aided ULSI design towards 21st Century	Prof. Y. Kajitani
THOMAS	Design method for flexible manufacturing structure of global network type	Prof. Y. Ito

Table 4 Subjects in Creative Research Project of T.I.Tech.

project under acronym of "THOMAS" (refer to preceding Chapter) is in directly close relation to the machine tool science and engineering. Fig. 8 and Table 5 show the outlines of this research body of the present, i.e., organisational structure, finances, research subjects and so on.

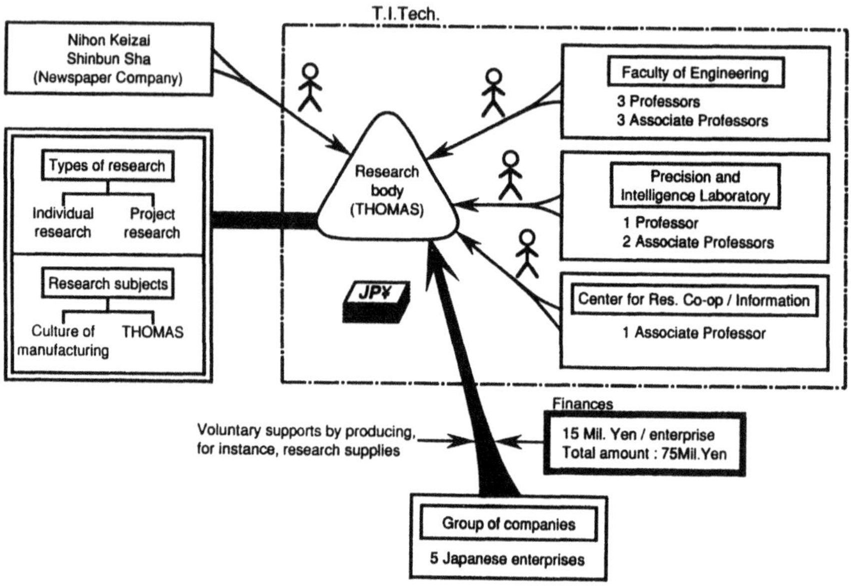

Fig. 8 Outlines of Research Body "THOMAS"

Types of research	Research subjects	
Project research	Culture of manufacturing	Production technology for region-specified goods
	THOMAS	Formulation of production-related deep knowledge
Individual research	Culture of manufacturing	Production-related infrastructure within NIES / desirable structure for international co-operation Feasibility of science park as to be a base for technology transfer
	THOMAS	CAD system for mechanism based on the intermediation of sensitivity with logic Virtual assembly system Work analysis of skilled technician / model of craft learning process Expert systems for solving vibration and noise problem In-process sensitivity sensing

Table 5 Details of Research Subjects within THOMAS Activity

As can be seen, the project can be characterised by its germinating and engendering research works, which are, of course, very attractive to the university staffs. Although remaining something to be seen, in contrast, the company can expect to have got some advantages, e.g., "Seed-Oriented Information", which could assist to draw up the company's future strategy.

3.5 University's Own Research Activities

In this case, the research works are, in principle, very basic and carried out with university's own finances and scientific research fund of the Ministry of Education. Of course, there are no obligation to determine the research subject, and thus the academic staff can settle down to her or his own subject in accordance with her or his interest and also in consideration of the international state-of-the-art of the interesting field.

By the way, we must mind that there are similar research works conducted by the staffs of the national laboratories, enterprises and others. These research works are regarded as to be another kind of university's own research activities.

Generally, the out-puts of these research activities are publicised through the presentation of the paper at the regular or irregular meeting, and also at the conference of the professional societies. The machine tool manufacturers can use these results as they like and sometimes they can ask the academic staffs of the university and others about more details of the research result. In certain cases, a co-operative work could be launched out based on this first connection.

Figure 9 shows some typical research subjects reported on the occasion of the recent regular meetings of the professional society.

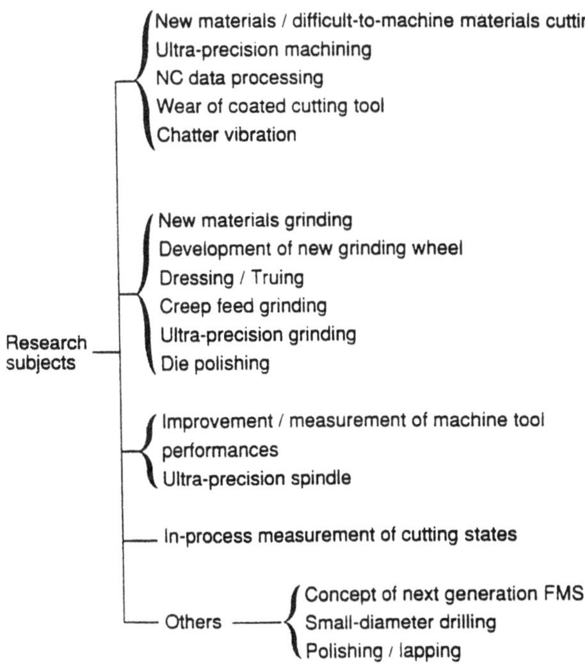

Fig. 9 Representative Subjects at Autumn Meeting of JSPE (in 1991)

As can be seen, the basic research has been carried out ranging from the detailed analysis of metal cutting mechanism to the proposal of the system concept. Generally, the qualities of these research works are very high; however, there is, dare to say, still one serious problem, i.e., "Remains of Europe and USA Worship". This induces the following trends in the Japanese machine tool research.

(1) A very few achievements on the innovative research work, proposal of creative concept, suggestion of new approach to the problem and so on. Fortunately, this weakest aspect is gradually improved, and Tables 6 and 7 enumerate some achievements, which are considered to be duly influential to the international society.

Table 6 : Some Leading-edge Technologies Recently Proposed and Developed by Japanese Researchers-Hardware.

Researcher(s) / Organisation	Achievements
Dr. Hisayoshi SATO / MEL of MITI	Multiple re-generative effects in self-excited chatter
Prof. Katumasa SAITO / Hokkaido Univ.	Die polishing transplanted technique of experienced worker
Prof. Takeo NAKAGAWA / Univ. of Tokyo	Multiple-laminated sheet configured die / Cast iron-bonded grinding wheel
Prof. Yoshio SAITO / Chiba Univ.	Entrained air and heat flows around rotating grinding wheel
Prof. Yoshimi ITO / Tokyo Institute of Tech.	Stiffness of bolted joint / Measurement of interface pressure by means of ultrasonic waves

Table 7 : Some Concepts and Innovative Ideas Recently Proposed and Developed by Japanese Researchers and Related People-Software.

Researcher(s) / Organisation	Achievements
Prof. Kazuaki IWATA / Osaka Univ.	Cell configuration based on task broker concept
Prof. Yoshimi ITO and Assoc. Prof. Hidenori SHINNO / Tokyo Institute of Technology	Structural description of machine tool / Design methodology for compactly cubic-like FMC (system machine) / Analysis of thinking way of mature engineer and its transplantation to computer environment
Prof. Tohru IHARA / Chuo Univ.	CAPP of flair type / Evaluation of difficult-to-machine in process planning
IROFA / JEIDA	Prediction of manufacturing environments in the year 2000

(2) Easy attitudes of following in USA's and Europe's step of the research activities so far performed, and also of choosing the research subject, which is highlighted and in popularity, but not concerned with the essential features of the due problem.

Nevertheless having such problems, these research activities accelerates indirectly the enhancement of the market competitiveness of the product. Table 8 shows the man power, finances and research subjects of author's laboratory.

Table 8 : Outlines of Research Unit within National University.

Man power		Academic staffs	Professor 1 / Associate Professor 1 Research associate 2
Man power		Postgraduate students	8
Man power		Undergraduate students (4th year)	5
Man power		Research students	2
Man power		Visiting researcher	1
Finances (Mil. Yen / Year)		Running cost (from Ministry of Education)	3.4
Finances (Mil. Yen / Year)		Scientific research fund (3 subjects / from Ministry of Education)	12.3
Finances (Mil. Yen / Year)	Research funds	From Professional society	3.0
Finances (Mil. Yen / Year)	Research funds	From semi-national body organisation	0.5
Finances (Mil. Yen / Year)	Research funds	From companies	3.0

Major research subjects	
Machine tool design / structure	Development of ultra-precision diamond turning machine and its components Ultra-precision diamond turning for titanium alloy Thermally preferable design for structural body
Sensing technology	In-process measurement for machining environment with heat flux sensor Sensing technology for unevenness of colour shade
Design methodology for machine tool and systems	Design methodology for FTL based on machine, form-generating movements and process descriptions Design methodology for cubic-like compact cell of mobile type
Thought model-based manufacturing	Flair-based CAPP Quantification for feeling of difficult-to-machine
Culture of manufacturing	Manufacturing structure for region-specified product Measures to enhance the market competitiveness in Japanese machine tool industry

This is a typical example of basic research unit within the national university of old fashion, i.e., old imperial university and technical institute established before the 2nd World War, and within our Institute we have now about 7 units directly related to the machine tool engineering.

4 MEASURES TO ENHANCE MARKET COMPETITIVENESS OF PRODUCTS IN MANUFACTURERS

The manufacturers have been and are conducting a lot of activities from various aspects to enhance the market competitiveness of the product. Actually, the manufacturer has an organisation, which is, more or less, in charge of the research and development of the new product and also of the technology improvement. It is, however, very difficult to have the exact knowledge about the due organisational structure, ongoing development subject, forerunning activities and so on, because of company's confidence. Thus, we can suspect what have been done within a company by investigating the new product. In other words, each company's due activities so far performed can be understood through the dimensional and performance specifications of the new product.

Figure 10 summarises some marked products recently on the market and new

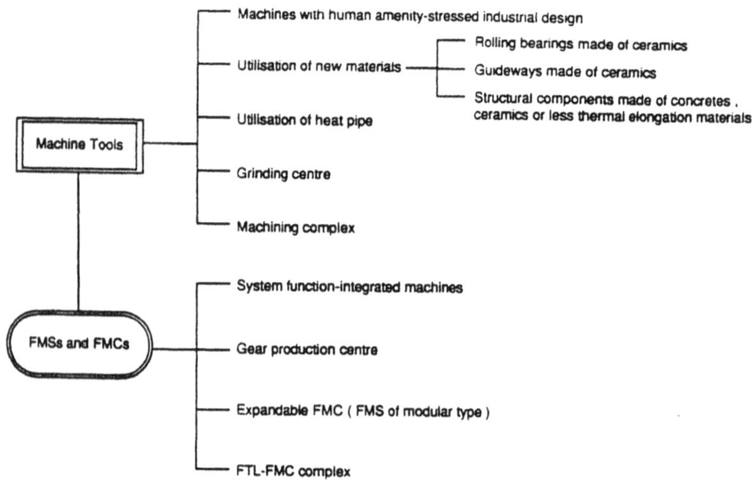

Fig. 10 Some Marked Products and Technologies Recently Employed by Manufacturer

technologies used by the manufacturer. From these evidences, it can be seen that the manufacturer has used various ideas and methods to enhance the market competitiveness, depending upon each company's philosophy and strategy.

The measures are ranged, for instance, from the concept design to the increase of the main spindle speed. For the sake of further understanding, Figs. 11 and 12 demonstrate such typical products[7].

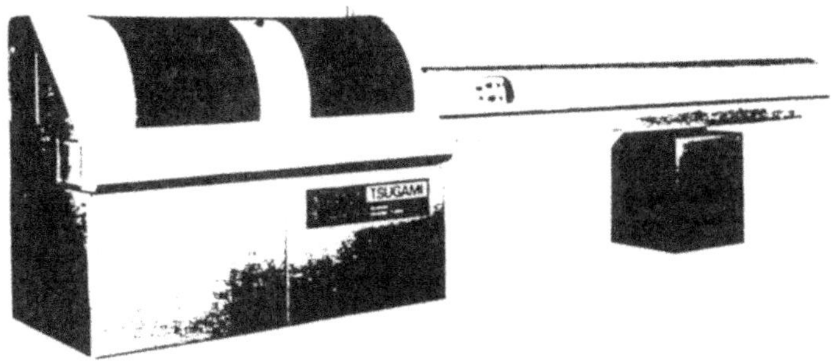

Fig. 11 CNC Automatic Turning Machine in Consideration of Human Amenity (Model: S20/Tsugami Co.)

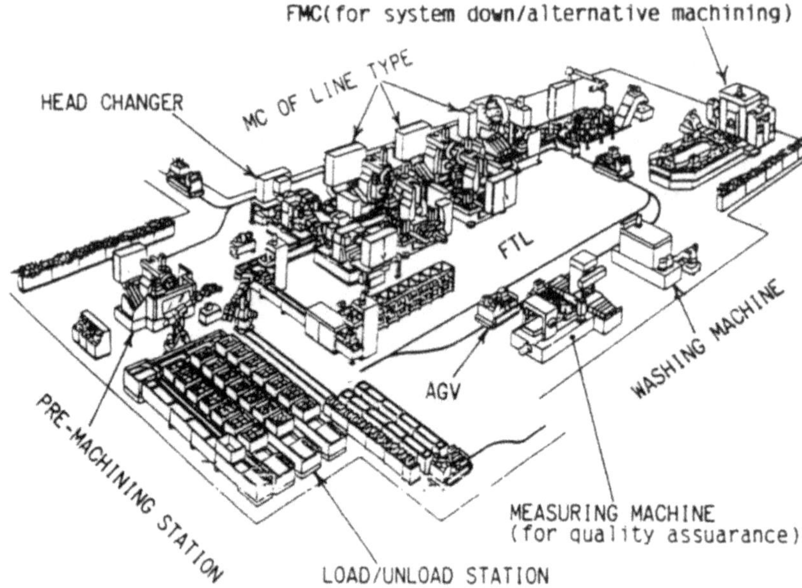

Fig. 12 FTL-FMC Complex (Model: FCIMS/Toyoda Machine Tool Mfg. Co.)

In the case of the machine shown in Fig. 11, the characteristic feature is to positively employ the industrial design to provide the operator with the amenity-related impression or feeling, i.e., enhancement of the marketability from the human factors. In contrast, the system shown in Fig. 12 is an amalgamation of the FTL (Flexible Transfer Line) with FMC (Flexible Manufacturing Cell), i.e., FTL-FMC Complex. As can be easily understood, this amalgamation requires of the very skillful design ability, because the system characteristics of the FTL are quite different from those of FMC, although these two systems belong to the same category of the flexible manufacturing. Briefly saying, this example can be interpreted as to be the enhancement of the marketability from the system performance for the high operability.

To deepen the argument, the two questionnaire investigations have been furthermore carried out, focussing on the following issues.

(1) The procedure and duly important information when determining the design specification of the new or major products especially from viewpoint of enhancing the market competitiveness.

(2) Noteworthy measures and tools in design and manufacturing procedures to enhance the market competitiveness of the product.

In these questionnaires, nine manufacturers have been chosen in consideration of the kind of products, production volume, technological qualification and so on. As a result, a lot of fruitful results could have been obtained and some of them will be stated in the following (details will be publicised elsewhere when having a proper opportunity).

Figure 13: Important Information in Determination of Specification of New or Major Products.

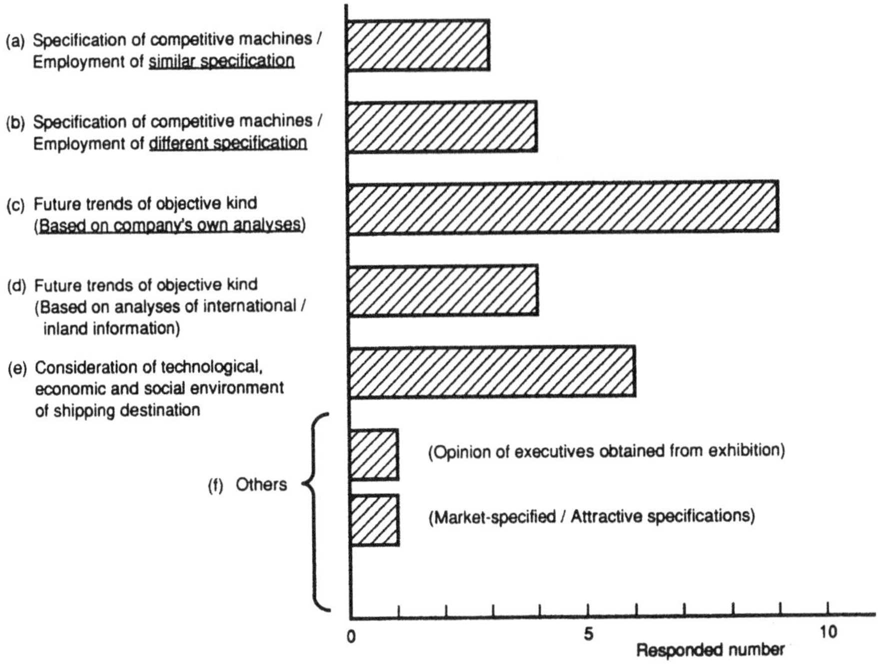

(1) Most of machine tool manufacturers determine, as shown in Fig. 13, the design specification of the new or major product by mainly attaching the importance to the following information.

(a) Future trends of the corresponding kind of machine tools.

(b) Consideration of technological, economic and social environment of shipping destination or expectable market.

When referring to this diagram, furthermore, we must be aware of it that the necessary information are duly varied depending upon the kind of machine and also on its production mode. For instance, there are the two typical cases as shown in Table 9.

Table 9: Variation of Necessary Information Depending on Kinds of Machine Tools.

Manufacturer / Kind of machine		Information attaching the importance	Remarks
A	Mass production type	(a) and (d)	Universal-purpose NC Lathe
	Special-purpose type	(c),(d) and (e)	Machine for non-circular cutting / High precision machine tool
B	Kinds having powerful competitiveness	(a)	
	Kinds coming belatedly into the market / Kinds having less competitiveness	(b)	

Table 10: Measures Commonly Employed by Many Manufacturers to Enhance Market Competitiveness.

<table>
<tr><th colspan="2">Procedure</th><th>Measures</th></tr>
<tr><td colspan="2">Market surveying</td><td>Correct and timely acquisition of market information and user's requirements</td></tr>
<tr><td colspan="2">Design</td><td rowspan="3">Reduction of production cost /
Production of parts and sub-assembly with higher quality</td></tr>
<tr><td rowspan="3">Manufacture</td><td>Machining</td></tr>
<tr><td>Assembly</td></tr>
<tr><td>Inspection</td><td rowspan="2">Complete inspection or shortening of inspection time by improving parts and sub-assembly qualities in machining and assembly procedures</td></tr>
<tr><td colspan="2">Product inspection</td></tr>
<tr><td colspan="2">After sale service</td><td>Quick servicing system - e.g., Deployment of service bases /
Regularly travelling maintenance service</td></tr>
</table>

(2) Although standing on the different environment from one another, as shown in Table 10, the manufacturers employ roughly the common measures and tools notwithstanding the related procedures (see Fig. 14) in the production to enhance the market competitiveness, e.g., timely collection of market information and reinforcement of after sale service. In addition, it is worth pointing out that the details of each means are, of course, varied depending upon the technology available at each manufacturer. For instance, the reduction of the production cost is tried mainly at the design, machining or assembly procedure using the unit construction, reduction of work-in-progress and shortening of lead time.

Figure 14 shows, furthermore, some representative measures and tools being employed in the Japanese machine tool manufacturers of average level, wherein the Manufacturer A has mainly produced the universal purpose-oriented MC and TC, whereas the major product of the Manufacturer B is the large size, special purpose-oriented NC machine tools.

		Market surveying	Design	Manufacture: Machining	Manufacture: Assembly	Manufacture: Inspection	Product inspection	After sale service
Manufacturer A	Measures	Consideration of business requirement of every shipping destination	Comparison of other companies' specifications / Standardisation	High accuracy machining / Realisation of reasonable cost	Less parts inventory / Unit construction to respond customer's various requirements	In-house inspection system	Less troubles and malfunctions (teething troubles) at initial stage operation	Quick servicing system
	Tools / Methodology	Questionnaire investigation / Product planning to be reflected user's requirement	Know-how accumulation of due technology / Assessment test of units	FMS for parts machining			Final running test before shipment	Quick supply and control system for spare parts
Manufacturer B	Measures	Interview to users / Related data collection	Proposal of new concept / Co-operation with user in full consideration of user's production technology	Employment of new cutting tools to establish high accuracy and performance machining / Accuracy control by temperature-constant environment	Shortening of lead time	Full inspection of all the parts	Running test	Devotional regular maintenance system
	Tools / Methodology	Machine selection system considering the kind of users			Modular design / Use of temperature-constant factory / Uniformalization of parts quality by controlling machining data	Co-ordinate measuring machine / Data feed-back to design and manufacturing	Excitation and cutting tests	User's data control by computer / Consultancy of operational technology / Trouble shooting committee

Fig.14 Some Examples of Measures and Tools in Production Procedure to Enhance Market Competitiveness

When discussing the company's activity, here, the one-to-one co-operation between the university and company should be noted. This co-operative structure is sometimes in the form of the contract research, and aims at the establishment of the remedies to a specified technical problem to be urgently solved. In due course, the co-operative work is carried out based on both the due technological resources accumulated so far in the university and obtained from the factory floor experiences.

In addition, we must be aware of it that the Japanese machine tool manufacturer has been very interested in the international deployment of the manufacturing and product development bases to reinforce the international competitiveness. On this extension, the region-specified product is now in the focus of the argument from the viewpoint of its marketability. This paper has not touched with it in deep; however, the necessity is to understand more details of this way to enhance the marketability. In fact, one of the sister companies of Makino Milling Machine Co. in Singapore is producing the region-oriented MC and this machine is said to have strong competitiveness within that area.

5 CONCLUDING REMARKS

It is generally very tough works to reveal the overall view of the research and development activities, which have been and are being carried out within a nation, although limiting the objective industry. It is believed that this review paper is a first trial to explicate the Japanese structure and system to enhance the marketability of the machine tool from the technological aspect.

The paper appears to be useful; however, it is imperative that the following investigations be, for instance, carried out hereafter to have more insightful knowledge.

(1) Entity of the voluntary supports to the university by the enterprise. This has a considerable effect on the research quality. For instance, the university can purchase the research supplies with less than the turn-over--value.

(2) The role of the research and development within an enterprise to enhance the marketability of the product.

Although including insufficient explications, author is convinced, this paper will be somewhat helpful for those, who are interested in the analysis of the marketability of Japanese machine tools.

REFERENCES

[1] For example, Ito Y. "Conceptualizing the Future Factory System". Manufacturing Review of ASME 1988; 1-4 : 252-258.

[2] Ito Y. Amalgamation of Human Intelligence with Highly Automatised Systems - An Approach to Manufacturing Structure for the 21st Century in Japan. In: Olling GJ and Kimura F (eds) Proc. of IFIP TC5/WG5.3 8th Inter. PROLAMAT Conf. - Human Aspects in Computer Integrated Manufacturing, Elsevier Science Publishers B.V., Amsterdam, 1992, pp 11-26.

[3] Ito Y, Shinno H, Nakanishi S. "Designer's Thinking Pattern in the Basic Layout Design Procedure of Machine Tool - Validity Evaluation of Thinking Block". Annals of CIRP 1989; 38-1 : 141-144.

[4] Ito Y. "Flexible Manufacturing System Complex Provided with Laser - Part 3 Application System Design". In: Proc. of 5th ICPE, JSPE, Tokyo, 1984, pp 28-36.

[5] Ito Y. "A Desirable Production Structure Looking Toward the 21st Century - Anthropocentric Intelligence-Based Manufacturing". In: Proc. of 11th ABCM Mechanical Engineering Conf., Brazilian Society of Mechanical Sciences (ABCM), Sao Paulo, 1991, pp 23-32.

[6] Moriwaki T, et al. "Grundlagen-und Anwendungs-Forschung über Bauelemente aus Polymerbeton". Industrie-Anzeiger 1983; 105-17 : 48-53.

[7] Ito Y. "Recent and Near Future Trend of Japanese Machine Tools and Related Technologies". In: Handouts for Overseas Technical Seminar, JMTBA/MIER, Kuala Lumpur, 1991, pp 1-31.

Synthetic, Pragmatic, Analytic - A Comparison of the Japanese, American and German Approaches to Machine Tool Design

by Eckehard Fozzy Moritz

1 Preamble - Mission, Means, and Methods

Good news! "Culture" is on the rise again! Whereas a mere decade ago computer integration, automation, and, in areas not (yet) to be automated, strictly supervised project management were praised as the formidable chance to reduce troublesome and error-prone human process interference to a negligible residual parameter and to advance into a bright future of opposition-free production and profit maximation, in the last few years the strict limitations of these approaches have, sometimes quite painfully, been demonstrated. Nowadays no serious discussion about the future of industries or the improvement of management practices can continue to neglect the "human factor" - neither in its individuality nor in work organization - or other culturally influenced parameters as the structural integration of the company, the historical tradition of the industry, or the social environment of the country.

Even the academic world has recognized this trend. Scholars from different disciplines are unanimously beginning to realize that a truly integrated and interdisciplinary view is essential to approach a real understanding of the functioning of something as complex as an industry. And in this context culture can no longer be confined to "explaining" the inevitable uncertainties (yet) left in an otherwise purely mechanistic conception of the world. Just the reverse is true; eventually culture must be seen as a pivotal constituent of **any** action happening within **any** social system, thus also within industries, thus also within a company.

To account for this new interconnected understanding of industrial practice and processes the term "industrial culture" has been coined. And, as was indicated above, rather than simply supplementing existing deterministic ways of describing technical and societal correlations by including a few cultural parameters, this terminology aims at stimulating a whole new discussion reconsidering how industries work, in what way(s) cultural factors influence industrial phenomena, and, last but not least, how cultures themselves function.

The first few chapters of this book laid a thorough theoretical groundwork for an understanding of the concept of industrial culture. However, this deductive approach is not the only possible technique to proceed with this complex venture. Relying on analytical conceptualizations alone should rather be regarded with some suspicion, as again important factors and environmental parameters determining or influencing industrial practice may be neglected.

Therefore, in this paper I shall try to approach this problem from an empirical perspective. This is not only a good probing of the analytical considerations elaborated in other contributions to this volume, but will hopefully also help to supplement the existing models and thus lead to a better and especially more comprehensive understanding of the multifarious facets of industrial culture. As an intended side-effect, this contribution also wants to put some of the recent single-tracked claptrap studies, notably the "lean" discussion, into perspective.

The last few paragraphs outlined the basic objective of this paper. However, the sheer complexity of this task just does not permit an adequate representation and discussion of all potentially interesting aspects of industrial practice and their relevance to this topic. Therefore a number of limitations of this contribution must be straightened out beforehand.

First, the scope of this research project was restricted to analyzing the product development procedure in the machine tool industry. Thus, although a few more general statements were included to illustrate some important country-specific social or intellectual characteristics, most of the results and conclusions of this contribution must be interpreted as to apply only to this specific area of research.

A second important point to be acknowledged is the somewhat unfortunate fact that this paper includes a lot of generalizations, assertions and contentions, and positively too little differentiation and discussion of the respective backgrounds. However, this does in no way mean that there has been no such reflections or distinguishing considerations. It was just a question of deciding, when conceptualizing this contribution, between sticking to a few points and analyzing these into some detail, as is mostly done in scientific papers, or attempting to illustrate a great number of perspectives simultaneously influencing one and the same practical industrial process, at the cost of in-depth analytical confirmation. And in accordance with the fact that a more integrated understanding of industrial processes is what I consider most important in advancing an understanding of the notion of industrial culture, I chose the latter.

Nevertheless, as this contribution summarizes some of the most important conclusions of my doctoral dissertation, you may be assured that the contents are derived from a broader analytical as well as empirical basis as the representation of results and conclusions here suggests. To at least give an idea about the extent of the empirical work done (that also constitutes the basis of this contribution) I will briefly summarize the methodological approach of the overall investigation.

The groundwork of the whole research project was formed by comprehensive interviews with senior designers or design managers of at least eight machine tool companies each in Japan, the United States, and Germany. These interviews were conducted in the native language of the interviewees - not only methodologically, but also pragmatically a necessity - and were complemented with factory tours including a visit to the design department and the shop floor. Two accompanying questionnaires, a few intensive case studies in each country lasting from a number of days to two weeks, a constant mutual communication with some selected companies as well as with experts of various fields, and an extensive literature search completed the empirical approach of this investigation.

2 Introduction - the Protagonists and their Companies

Before discussing the most interesting results of this investigation, a few preliminary considerations should be dedicated to a brief introduction of the interviewees and their companies. This seems necessary for two reasons.

First, understanding the profile of the companies and other facets of the industrial order does not only help to get some more insight into the functioning of the machine tool business in each country, these factors also exert quite some influence on the shaping of the respective product development patterns. Therefore, the selection of the companies aimed at mirroring a representative sample in each country as far as the products, the size of the companies, the batch sizes, and the geographical location within the country is concerned; illustrating some of these aspects is then a solid groundwork to prepare and improve an understanding of the following discussion.

Second, acknowledging the importance of the interrelatedness of the different constituents of industry is in no way confined to the company level alone. Design managers as well, though apparently some of them are not likely to agree, cannot be regarded as technically brilliant skulls isolated in space; their way of thinking and acting is also shaped by their educational and professional background and is constantly influenced by their technological, industrial, and intellectual environment. It is likewise interesting and necessary to examine at least their professional backdrop into some more detail.

Figure 1 illustrates the distribution of the respective size of the companies measured in the number of employees. It may surprise some of the readers that the differences in the average size are almost negligible considering the often stated assertion that Japanese companies are simply much bigger than their counterparts overseas. But although this may be valid for the huge mutually intertwined conglomerates, and seems to be suggested by the turnover statistics in the American Machinist year for year reserving at least seven of the top ten spots for Japanese companies, as far as the number of employees is concerned this assumption does not hold true any more.

The Japanese industrial structure of course has its own distinctive features. There are comparatively few of the fine breed of small special machine makers which are the pride of the German machine tool industry. There also exist a

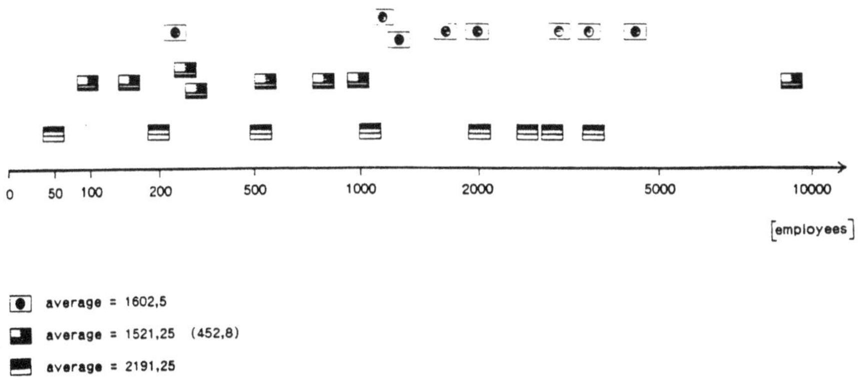

Figure 1. The distribution of the number of employees of the companies visited

number of obviously successful and, regarding the relationship of the number of production workers to the turnover and profit margins, supposedly very efficient big machine tool companies which attract the attention of competitors and researchers alike. Nevertheless, even those headliners of international comparative statistics do not host considerably more wage earners than their counterparts in the other countries of this investigation. The fact that a higher percentage of Japanese as compared to German or American machine tool makers fall into the category of 100 to 500 employees (e.g.[2]), and that their performance is on average just as mediocre as that of any other industrialized nation's enterprise, seems to go unnoticed altogether.

In the case of the United States, however, the mean is really exaggeratedly high regarding the collapse of the American machine tool industry in the last decade. But this phenomenon can easily be traced to one single reason: one company, which has become close upon being synonymous with machine tool making in the United States, is distributed all over the country and engaged in the production of almost all kinds of machinery. Its still huge overall size is the sole reason behind the distorted statistics; leaving out this firm reduces the average to a much more sensible value.

Figure 2 summarizes some of the personal data about the interviewees. Here as well a few facts are worth noting:

The Japanese profile is clearly the most balanced - the higher the rank*, the older the person. This is not very surprising, though. The seniority principle,

---

* It is important to take into account the overall size of the company as well, especially in the higher management positions. The director of engineering in a company employing 3500 people must be regarded as of "higher" rank than the same position in a company with 100 wage earners.

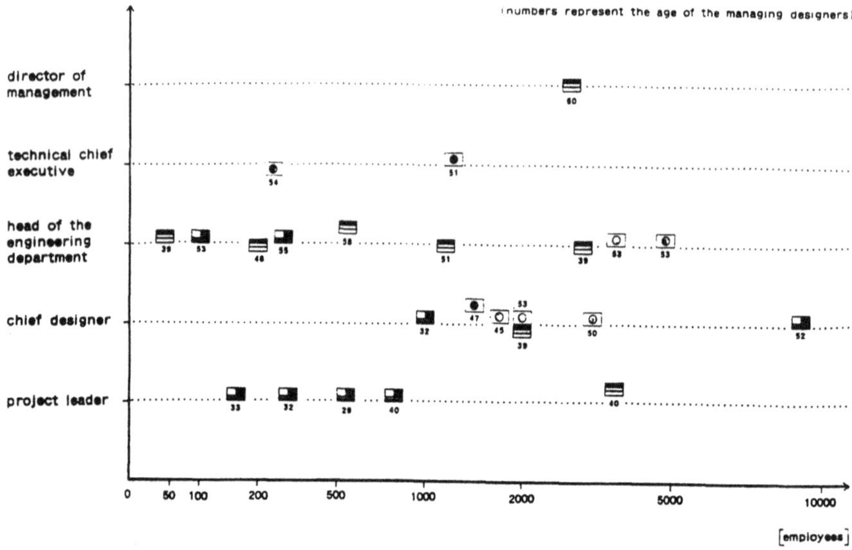

Figure 2. The hierarchical position of the interviewees

career advancement by age (at least in a formal respect), is still one of the important pillars of the Japanese personnel management system, which obviously also reflects in the positions of the design engineers. However, runaways are possible, as one "just" 51-year-old chief executive, well known in this industrial sector for his eccentricity and brilliance, demonstrates.

In Germany the distribution of ranks and positions is more uneven. There are many ways of becoming a design manager, from the rather traditional step-by-step advancement of perceptive draughtsmen especially in smaller companies to the presently fashionable policy of providing diploma engineers from technical universities a shortcut to positions in the design management within a few years. The two 39-year-old directors of the design department are clearly of this breed.

In the United States the situation is most difficult to comment. The "hire and fire" strategy of American firms has fully shown its disastrous effects on the pitiful machine tool industry, a few companies having laid off as many as 80% of their employees within the last ten years. Thus the profile of the interviewees constitutes a mixture of a few old, experienced chief designers, some leftovers of the "leaning" process, and a few brilliant and ambitious young designers in less poorly performing enterprises. It is important to note, finally, that the average position and age of the interviewees is comparatively low; besides the American pragmatism another reason why a number of the answers to be discussed in the following chapters are closer to "practice" than in either Japan or Germany.

(engineering) education	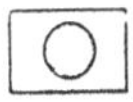		
general assessment of the engineering education at the university	- diversified - a lot of general education - little specific knowledge	→ like Japan	- very specialized - intends to induce "engineering thinking" - might produce narrow specialists
emphasis of design education	- adaptation } of an existing design - optimization } of an existing design	design project	creation of a conceptual solution (new concept design)
education in design methodology	product specific	conveying a procedure	problem solving oriented
(further) education inside the company	- freshman year - OJT (on-the-job-training)	- little - specific	- increasing - seminars and training programs

Table 1. Some characteristic features of the engineering education in Japan, the US, and Germany

3 Education and Career - Moulding the Masters of Engineering

Table 1 summarizes a few interesting facts about the differences in the engineering education in Japan, the United States, and Germany. This compilation shows the results of a former research project [3], and is complemented with the experience I have gained being part of the "system" (as both student and assistant in Japan and Germany, and as professor in the United States) within the last few years.

An obvious yet interesting observation is that the engineering education already introduces a number of characteristics into the maturing engineer which may be traced like a central thread in all kinds of different activities and proceedings throughout his professional career. The most prominent of these influences is the trend towards specialization and analytic thinking in Germany which, as will be shown later, results in an attitude towards designing and work organization which in many aspects differs from the more generalized and pragmatic inclination of the Japanese and American engineers.

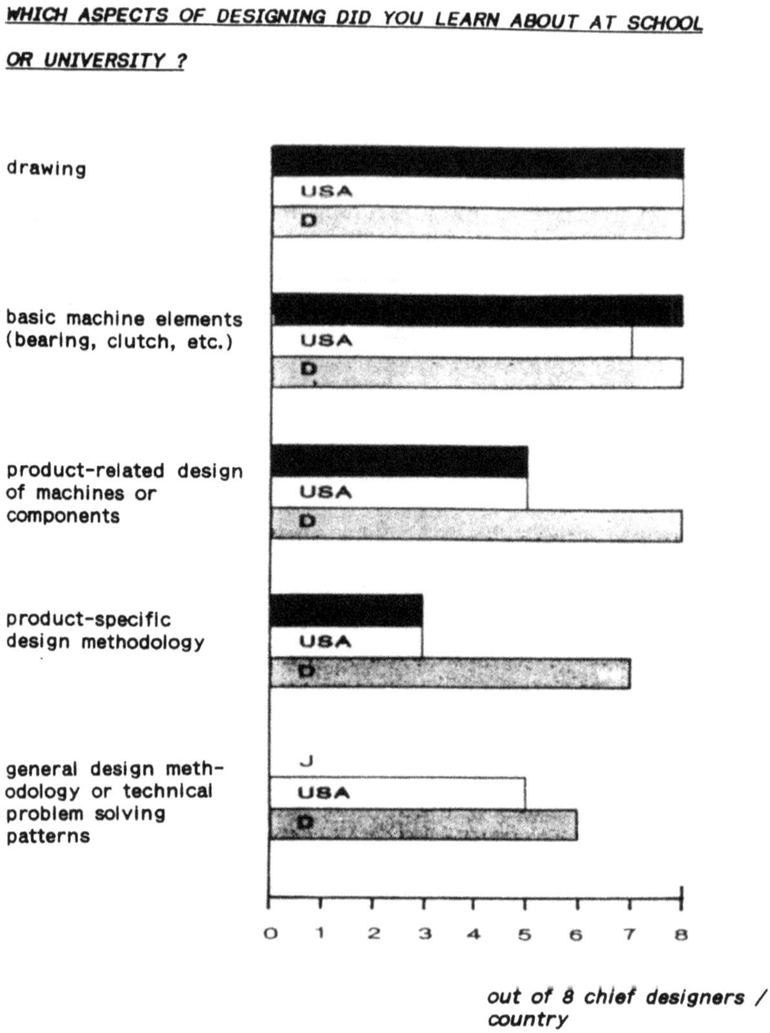

Figure 3. Some (design-)relevant aspects of the educational backgrounds of the interviewees

The actual contents of the design education are sorted out in Figure 3. The main message of this figure is the fact that a German engineering education, while quite specific and prone to produce narrow specialists, on the other hand certainly constitutes a most valuable academic foundation for maturing designers - surely one reason for their technical excellence.

career			
change of companies	very rare	rare	increasing with augmented hierarchical level
comparatively "uncommon" career stages of chief designers	- patent department - product planning department	research department	no job outside of design departments, rather design of a different product
potential range of occupations	diversified, inside a company	technical, within one industrial sector	restricted to one's special field, (almost) any industrial sector

Table 2. Some characteristic features of career patterns in Japan, the US, and Germany

Table 2 summarizes some interesting facts about the three distinct ways to become a design manager. The most interesting finding here is the versatile profile of the Japanese. Not that there is variation at all; the Japanese fondness of job rotation has been widely discussed in literature, and by nowadays this phenomenon may even be considered a rather disappearing peculiarity of the Japanese personnel policy. But one of the at least not untypical stages in the career of Japanese design managers, a position in the product planning department, hardly even exists in either American or German machine tool companies; another one, the occupation as a patent engineer, is "usually" considered a highly specialized job. One could hardly imagine a patent agent becoming a chief designer later on in his career anywhere else in the industrialized world but in Japan.

4 The Relationship between Industry and Culture - Overwrought and Neglected at the same time?

This apparently contradictory headline intends to provoke and shake up at the same time. Cultural, social, and historical factors influencing industrial practice and performance are widely referred to as important and decisive elements in accounting for national peculiarities in manufacturing and production styles. Reviewing those "explanations" on a more specific level, however, reveals that those who spread the mystical umbrella of the "Meister"-system, the boundless freedom of the individual, or "lean production" above the different production systems more often than not have little to say about just how these phenomena actually function in practice and why exactly they are that much superior as opposed to other forms of manufacturing organization. On the other hand, managers in industry or engineers who concern themselves with the practical details of production usually sternly avoid even considering the possible impact of "soft" social or cultural peculiarities.

Thus it is one important goal of this paper to promote more communication among these two groups - for the sake of improving working conditions as well as of an increase in turnover figures, profit margins, technical excellence, or whatever else a company may be striving for.

"Industriekultur"	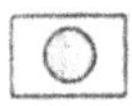		
economic system	feudal market economy	free enterprise	social market economy
relationship employer - employee (industrial relations)	holistic utilization of the employees	individual modes of regulation (contracts)	individual and social modes of regulation (within the framework of the wage agreement in the respective industrial sectors)
motivation of the employees	- sense of duty - promotion of (actually quite restricted) participative management - payment, capitalistic ideals recently increasing	- payment (wages) - subconscious approval of capitalistic structures	- payment (wages) - professional ethos
main objectives of industrial and economic policy	increase of - power - market share	increase of - money - influence	- technical and technological leadership - self-fulfillment (ego-boost)

Table 3. Some characteristic features of "Industriekultur" in Japan, the US, and Germany

Table 3 reviews some of the more "down-to-earth" categories, derived from empirical observations, that strongly influence and determine an industrial culture. This compilation is much more simplified and generalized than most people would deem permissible; however, there is hardly any practible alternative if these undoubtedly important characterizations of the industrial environment are not to be omitted altogether in this abridged report.

Among the bulk of explanations and comments that would actually be necessary to qualify this table, just a few of the most relevant and important ones can be broached here.

The feudal market economy postulated for Japan is certainly the most remarkable entry. This expression, while having been aired a few times already, in no way fits into the scientific categories of economics studies. But it is definitely important to acknowledge the oft-neglected fact that the economic system in Japan, while officially being named "free enterprise" just as in the United States, has certain features that strongly distinguish it from the market economy of its biggest trading partner.

Feudalism in this context should not be interpreted as having to do with aristocracy or nobility in the true sense of the word. Rather, it must be understood as a picturesque description of a mode of economic organisation in which, on the one hand, a few leaders of big companies and conglomerates play the game of capitalism with great strategic brilliance and tenacity while, on the other hand, the laymen are tied up in an all-embracing set of "universal" codes and rules of behavior which provides a basic form of social security and possibilities for self-identification but which at the same time allows a holistic utilization of the work-force.

Brain-washing motivation of the work-force plays a big part in this Japanese industrial policy. Sure enough, a motivated employee will definitely be more productive for the company and may also derive more satisfaction for himself from what he is doing anyway. On the other hand, however, German scholars may with good reason stress the point that self-reflection, perceptiveness, and a critical mind are the ultimate ingredients of individual intellectual emancipation; factors which can certainly not be attributed to the Japanese way of organization.

Still, in Germany the virtue of criticism frequently seems to be confused with the justification and cultivation of objection. More often than not, especially in bigger companies, constructive criticism lags way behind destructive disputes. And it is certainly not without reason that the German people have "won" the unofficial title of the world's biggest complainers.

But despite all that, the actual situation in the German industry is not that bad, neither for the employers nor for the work-force. "Social modes of regulation", this rather abstract entry (still Table 3) implies that in German companies it is just about as difficult as it is in Japan to lay off redundant employees. Nobody guarantees "lifetime employment" here, but besides the fact that this idyll can be enjoyed by only about 30% of the Japanese employees, Japanese personnel managers also find ways to get rid of these by for example "asking them to leave" or by downgrading their position if the economic situation in their company really leaves them no choice. Furthermore, German companies as well will have to find socially acceptable ways if they intend to reduce their work-force, they will have to take care for their former employees' immediate social security, and they will not deliberately get rid of their generally highly trained workmen, anyway.

company structure			
particularities and emphasis in the company organisation	- product planning department - market research / strategic planning - experimental department - patent department	- business management - research department	- design department - "basic design" department (Grundlagen-konstruktion)
main momentum for innovations	product planning	management	design

Table 4. Some characteristic features of the company structures in Japan, the US, and Germany

5 Company Organization - Reflecting or Originating Differences in the Overall Production Process?

Table 4 highlights a few specific features of the company structure and the organizational distribution of the product development process. Scholars of various disciplines still debate over whether the particularities of the respective company organization originate most of the distinctive features of the production process, or whether on the other hand strategic considerations regarding the allocation of work determine the formation of corresponding departments.

With respect to the results of the interviews and case studies this conflict appears to call for country-specific rather than universal lines of argumentation. In Japan the ability to manipulate complex structures to approach strategic goals pragmatically (to be more comprehensively discussed later) also reflects in the flexibility of the company organization. Redistributing manpower, reshuffling departments, and fading out responsibilities into the never-never of endless hierarchical cascades are an everyday matter of course inside Japanese companies. This almost instant adaptability to pragmatic needs is further supported by an inscrutable wages policy [4,5], and the general and non-distinctive educational basis of the employees discussed above.

In Germany established structures inside a company are much more rigid and thus more difficult to change or adapt. Part of the story is the professional ethos cultivated and cherished by the workers, educational institutions, and unions alike. The argument is simple - if you invest a lot of time and energy into becoming a proficient specialist, and if you take pride in your skills and in your job contents, then you will hardly be willing to be transferred to a job where your skills are no longer needed and where you become an interchangeable "salaryman" rather than the adored expert. However, if management is clever enough to consider these interconnections and to meaningfully involve the employees in the restructuring of the firm, it can usually count on an interested and professionally adaptable work-force.

American companies are no less rigid than most of their German counterparts, though for different reasons. A very important issue in work organisation, especially on the shop floor, appears to be the preservation of claims and negotiated rights, not much different from the fight for physical claims in the gold digging periods in the previous century. Business management seems to have gained a much better and comprehensive control over what goes on in the different departments of the company [6,7], and as a result industrial relationships have become to be dominated by sternly battled contracts and mutual agreements. This surely constitutes a reliable basis for the work-force as well as the management and is also "fair" in a certain sense of the word; however, the practical necessities and the job contents often seem to be forgotten in this rather statutory process.

As a result, in Germany as well as in the United States the structure of the company and the organisation of work are much less pragmatically oriented towards actual needs and intended strategies, and are more a reflection of attitudes, professions, and the accrual and protection of spheres of influence as opposed to Japanese firms. This context is crucial to an understanding of the distinctive features of company organisation and production strategies. It furthermore has to be taken into account, but is unfortunately still widely neglected, in the current discussion about the transferability of management practices.

6 Product Development and Design - Originating Unique Technology or Materializing the new Money-Spinner?

Table 5 must be considered as the most important and original information this article is able to provide. Much more so than at any other stage of the production process, it is the initial few steps of the birth and development of the prospective

product development style	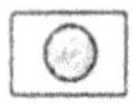		
general problem solving behavior	synthetic, eclectic	pragmatic	analytic, systematic
technical problem solving procedure	- experiments - accumulation of information	trial & error	- design - abstract reasoning
"control" of the product development process	- patterns - sequence of actions - "algorithmic"	- individual - according to the respective experience	- methodical - according to experience - "heuristic"
strong points in product innovation	- optimization - incremental improvements - instant transformation and application	eager to "check out something new"	- creative - systematic - "thorough"

Table 5. Some characteristic features of the product development styles in Japan, the US, and Germany

new product which are prone to be affected by a large number of internal and external, intellectual and social particularities of our cultural heritage. Nevertheless - or maybe due to this very fact - they have hardly been subject to any serious international comparative study so far.

The principal feature that distinguishes design and development from the later stages of the production process is that most of the action takes place in the brain, with paper and pencil just conveying the created information. The investigation and explanation of differences in the product development process cannot therefore be limited to technical or economic boundary conditions, phenomena in the industrial structure or work organization, or the level of education and qualification, as generally appears to be sufficient in the studies on shop floor labor. In examining distinctive features of the design behavior those nevertheless important factors still have to be complemented with more psychological or even philosophical qualities like the value-system, the self-conception, and the intellectual style, to name but a few.

One concrete and pivotal example of the far-reaching significance of differences in intellectual "styles" is the direct reflection of problem-solving techniques on the design procedure.

In the title of this report the Japanese way of coping with new challenges was paraphrased as "synthetic", the American style as "pragmatic", and the Germans were said to be "analytic". As these denominations are not necessarily self-explanatory, yet are vital for the further line of argumentation, they should be illustrated with a parable taken from day-to-day life:

A Japanese, an American, and a German intend to go from, say, Paris to London. How will they typically proceed?

The American will probably be the fastest. He will simply get to the airport and fly, not bothering which other means of transportation are available. The Japanese will more likely check his travel guide and ask a few acquaintances about their opinion, and he will then follow the most trustworthy or applicable way minutely. The German will accumulate time-tables, sort out train schedules and price lists, and will be eager to find the best possible way according to his present situation, all this while complaining that the Channel tunnel is not finished yet.

All three approaches - the pragmatic "go-ahead-and-do-it", the synthesizing of "ways" (the Japanese "do"), and the detailed analysis of this "problem" - will undoubtedly lead to London eventually. But it is the considerable differences in the proceeding that make this example interesting; and in more complex contexts it is no longer guaranteed that every way eventually leads to reach the intended goal.

How does this example translate to the product development process? As could easily be expected, the Japanese are absolutely superior in any kind of activity that precedes the actual design process, especially in the accumulation of potentially relevant information. This was not only one of the core results of a preceding study [8], but was also reinforced by the responses to a related question this time (Figure 4), which, unlike most of the other questions, was answered in great detail by the Japanese senior designers. And the reader should not be fooled by just four of them mentioning competitors' models as one source of reference information in this open question - this still sounds very much like the copying approach, an image the Japanese have had for a long time, and one which they are most eager to distance themselves from. There is no place in this world to attain a better overall view about Who sells What realizing Which principles regarding a certain merchandise than the product planning department of a Japanese company concerned with designing just this product.

Admittedly being good at "synthesizing information" still gives off more than a whiff of the imitation odour wafting around the Japanese engineering departments. But this is just one, however acceptable, negative view on this practice. In Japan itself there has never been anything wrong with duplicating, quite the opposite. According to Confucian philosophy, which at least until a few decades ago dominated the intellectual orientation of Japan, copying one's master is the utmost way of acquiring knowledge (and achieving benevolence).

FOR THE CONCEPT DESIGN YOU ARE CURRENTLY WORKING ON, WHAT DO YOU USE AS REFERENCE INFORMATION ?

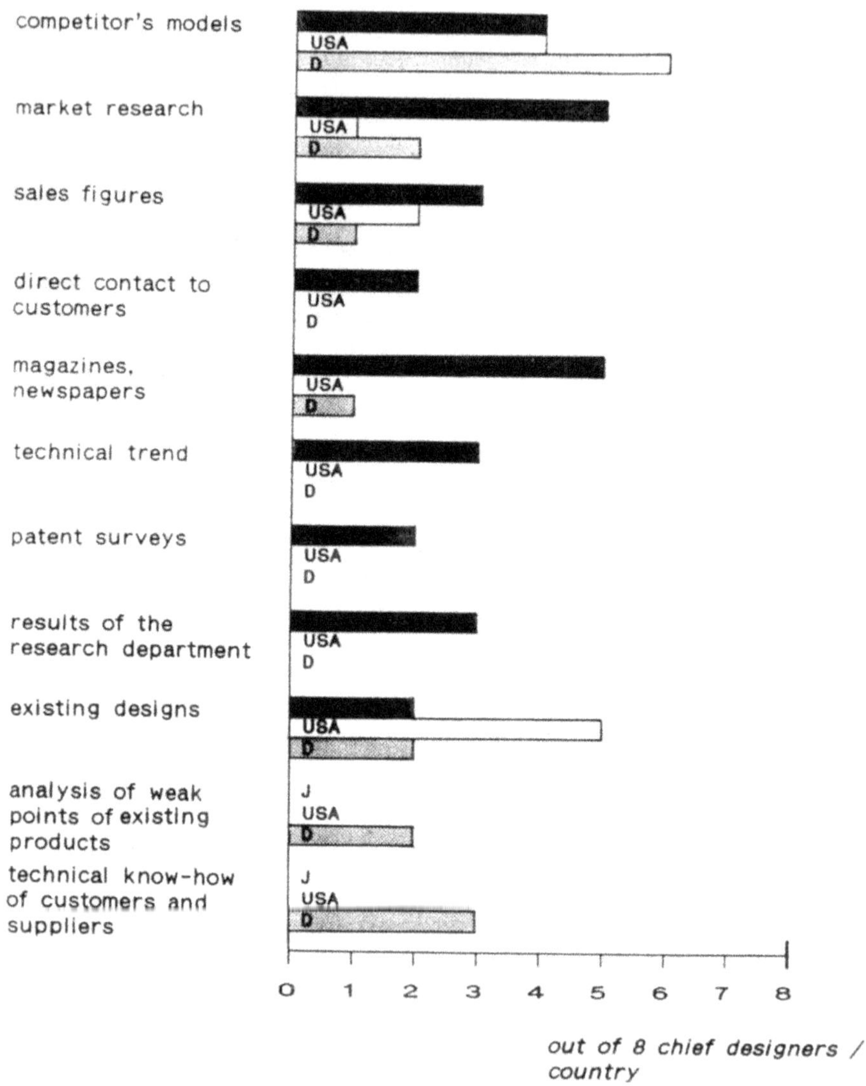

Figure 4. The input into product development - reference information used in Japan, the US, and Germany

And even without further indulging in philosophical speculations; reinventing the wheel due to an ignorance of the contemporary and available state of the arts in technology should rather be considered brainless than sublime behavior. Especially the almost exclusive mentioning of magazines, the technical trend, and patents as sources of information for prospective new developments on the part of the Japanese (still Figure 4) certainly illustrate a superior insight into the benefits of being broadly informed before attempting to innovate.

This is a book from which the American designers in particular should take a page out. Founding, as it seems, most of their product developments on existing designs gives a rather meagre outlook on the innovative potential of their companies. However, more than enough hints [6,7] point to the fact that it is the all-embracing might, preoccupation with "numbers", and conservativism of management rather than the lackluster intellect of the designers that prevent innovative breakthroughs in American machine tool companies.

Germany takes a kind of middling position in this discussion. Certainly the "analysis of weak points" is a highly recommendable thing to do and illustrates the beloved predilection of German designers for problem analyses. Nevertheless, the accumulation of information should definitely be devoted more attention than is currently done in German (and American) design departments.

The superiority of Japanese designers in accumulating information has another very practical origin. Creating new designs, especially surprising mechanical solutions to technical problems, can still not really be considered one of the strong points of the Japanese designers. Just as copying your master is highly esteemed in the Japanese society, inventing your own way is not; and it is just as little fostered by the conformity required in Japanese everyday life, by the rule-oriented structure of the Japanese intellect [9], or by the educational emphasis on "learning" instead of "understanding".

In the interviews this correlation reflected in the surprising dichotomy that the Japanese designers, when asked about the existence of product development patterns in their companies, answered in great detail and for the most part appeared to possess a pile of written documents for exactly this purpose. However, when asked more specifically, how they would proceed to design a feed drive for a new especially fast and precise prototype of a lathe, they offered just a few bits and pieces of hardly exciting ideas.

American designers answered most comprehensively to this very concrete question, offering more evidence that they are not necessarily to blame for the current lackluster performance of the American machine tool industry, and also furthering the hypothesis that their practice may be described as "pragmatic". But what exactly does pragmatic mean in this context? Henry Ford, possibly the most-cited figure in American industrial world, explained it this way: "We go forward without facts, and we learn the facts as we go along".

PROVIDED YOU SHOULD DESIGN A FEED DRIVE FOR A NEW ESPECIALLY FAST AND PRECISE PROTOTYPE OF A LATHE, HOW WOULD YOU PROCEED ?

Digest of the responses

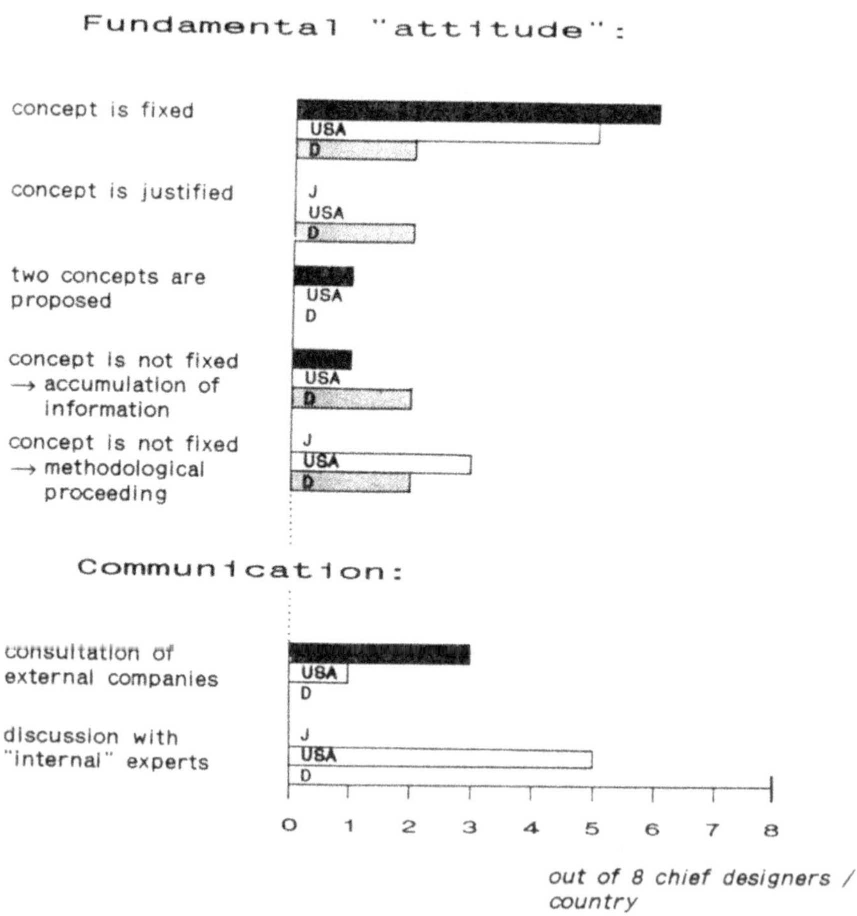

Figure 5. Some differences in concept design procedures in Japan, the US, and Germany

This statement, actually refering to his early success in the car business, is still astonishingly true, even nowadays, even in the machine tool industry. American designers typically choose one immediately interesting or at first glance promising way to reach a defined goal and pursue this way with great enthusiasm and insistence. They will not bother too much about a possible scientific backup or the satisfaction of having picked the "best" way, but they will usually be able to attain the intended goal quite quickly, and more often than not they will have surprising means of doing so.

Whereas the American designers devote most of their attention to reaching the goal, German engineers will more likely concern themselves with the understanding, formulation, and definition of the problem itself. The Austrian poet Grillparzer, although far from being an engineer, illustrated this behavior with an interesting metaphor: "If the Germans would have to decide between going to heaven right away or visiting a seminar about heaven beforehand, they would choose the latter."

Again, the "old" Grillparzer could hardly have portrayed the proceeding of a "modern" German machine tool designer better. The interviews as well as the results of a related prestudy [8] demonstrated that analyses of all kinds are a widely and readily acknowledged necessity in the proceeding of - almost exclusively - German designers. They want to fathom the problem comprehensively and from there on to concentrate on the potentially best action - a formidable means of developing technologically superior designs, but with the fatal probability of overshooting the mark or entirely losing sight of the intended goal.

Furthermore, as Figure 5 demonstrates, the German designers are least likely to take the technical state of the arts for granted. Just two of them would unconsciously rely on one "most popular" concept for a new design as opposed to five American and even six Japanese engineers.

Figure 6 indicates that the German design engineers also have a most comprehensive view of "their" product: a considerable part of the obstacles they were facing (and perceiving!) in concept design are complex technical problems.

The most important problem they are facing and do not seem to be aware of, however, is the lack of insight into the importance of communication. Surely German designers will ask suppliers, customers, or production people if they themselves consider it necessary. Often enough, however, they appear to see themselves as the only and all-embracing incarnation of technological wisdom. Whereas a number of Japanese in creating a new design would refer to external sources of information (an interesting result in itself!), and more than half of the American designers would get in touch with internal experts, none of the German designers mentioned any intention to contact anybody, neither within nor outside of his company (Figure 5).

WHAT WERE / ARE THE MAIN PROBLEMS IN THE CONCEPT DESIGN OF THE MACHINE YOU ARE CURRENTLY WORKING ON ?

Distribution of problem categories / country :

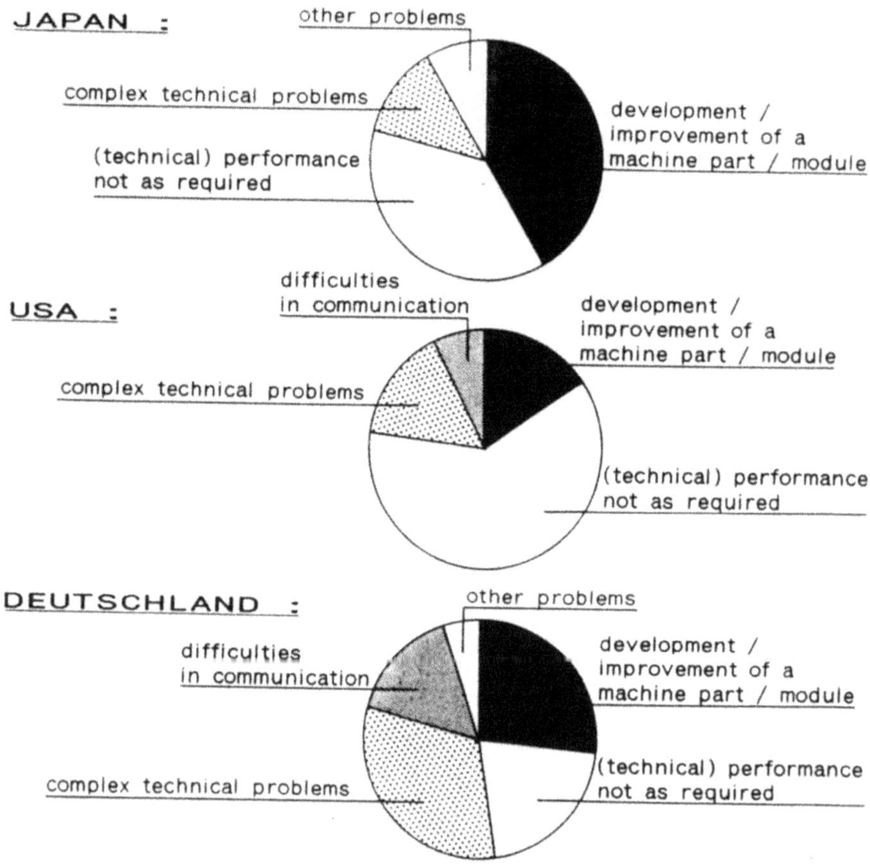

Figure 6. The distribution of (anticipated) problems in concept design in Japan, the US, and Germany.

WHAT DO YOU DO TO OVERCOME THOSE PROBLEMS ?

Categorization of the problem solving approaches :
(only multiple answers are listed)

JAPAN :

- optimization by experiments (4x)
- changeover to a different (known) technical principle and complementing experiments (3x)
- compilation of a matrix of possible solutions (2x)
- consultations with the shop-floor people (2x)

USA :

- formation of teams (3x)
- optimization by experiments (2x)
- changes in the design (2x)
- calculations (2x)
- "influencing the designer's way of thinking" (2x)

DEUTSCHLAND :

- changes in the design (4x)
- cooperation with specialists (3x)
- "influencing the designer's way of thinking" (2x)

Figure 7. Typical problem solving approaches in concept design in Japan, the US, and Germany.

One interesting and important point has not been touched upon yet. The Japanese above were decribed as being superior in the accumulation of information, but as having difficulties in concept design "from scratch". If that is true, how do they actually develop new machines or mechanisms where no "model" is available? The answer is made very obvious in Figure 7. Whereas in Germany the predominant way of handling problems in concept design is to actually change the design, the Japanese for the most part seem to resort to experiments. The abstract pondering and philosophying about the prospective performance of a yet two-dimensional drawing is definitely not their cup of tea, just as little as the German engineers are willing to extensively test all aspects of a mediocre design, and finally end up building something which may work but whose engineering or functional principles are not really understood.

Both approaches have advantages and disadvantages, and the decision about which one to employ should be based not only on the country one happens to work in, but also on one's individual working style, and, most important of all, on the kind of difficulties one is facing.

Generally speaking, unless the functional structure or the interrelation of influential parameters are very complex and thus neither an intellectual "grasp" of the overall problem nor its simplification is possible, changing the design is certainly cheaper and more "elegant" than experimental adaptation and optimization. Having "understood" the design does not guarantee that it really works, however, even if the designer may be very sure that it should. Functional misconceptions, the negligence of important parameters, or simply flaws in the anyway very empirical "theories" of engineering knowledge may and do offer continual surprises and problems in the realization of a new design.

This is a problem which the "experimentalists" do not have to bother with. Having achieved a design by experiment ensures by the very nature of the approach that the resulting mechanism will work.

The concluding recommendation is thus quite obvious. Generally the development and improvement of technical mechanisms by design should be the more efficient and beneficial method for new product development. The more complex and the less intellectually "graspable" a problem becomes, however, the more reasonable the experimental approach will be.

A second important parameter to be considered is the lot size. Whereas for small batch sizes experiments for purely financial reasons are usually not justified (unless it is a scientific or, unfortunately, military device), in larger lot sizes the development cost can be distributed and thus requires a much smaller share of the overall product price. Potential benefits of the experimental approach, such as the optimization of a design for manufacturing, may thus well pay off finally.

Two additional phenomena in Figure 7 still have to be discussed briefly. First, American designers, as the answers appear to reveal, will quite likely attempt to solve a technical problem by forming a team. Desirable as this may seem, it must be noted that at the time of the interviews Japanese product development strategies, which are said to be formidably based on team-work, were very much en vogue in the United States. Any mention of this aspect may therefore just as well be influenced by the current "scientific" discussion as by the actual emphasis on group-centered design approaches.

Secondly, the German answer to "contact a specialist", which after all was voiced three times, seems to contradict the above accusation that German designers neglect the aspect of communication in product development. However, this question concerned the actual measurements which were taken to tackle a problem. In the German industrial structure with its superb specialized suppliers, perceptive customers, and preponderance of custom-made designs, it is just a natural necessity to contact a specialist in the case of obstacles or problems.

Regrettably this does not seem to imply, however, that these specialists will be thought of and referred to beforehand in order to define important interfaces and to anticipate a number of problems before they may actually arise. Thus to allow a more integrated view and to foster communicative skills in German designers must still be recommended as an important goal for engineering education and management in Germany.

7 Conclusions

The presentation of empirical evidence and the related discussion so far left one most important question untouched: How, finally, may the different perspectives and the scattered in-depth reflections of this report contribute to fostering an understanding of the concept, the implications, and the actuality of industrial culture? In this respect, I would like to suggest six conclusions:

1. This article at least made very obvious that there is a huge and complex variety of mutually intertwined parameters and perspectives that define and determine an industrial culture.

2. This implicates that it still takes a lot of efforts to approach a truly integrated and comprehensive understanding of industrial culture(s). This contribution, and even this whole book, at best show the tip of the iceberg of this interesting as well as important area of research.

3. From a more functional point of view, this also implicates that the derivation of specific practical recommendations from this research to improve various aspects of procedures and management practice inside individual industrial cultures, although the objective as such is certainly reasonable and justified, has to be handled with quite some care - more care at least than was so far applied by most scholars in the international comparative field. In this respect, it is especially important to acknowledge the fact that changes of organizational features or management style inside an industry, although the outcome may at times prove to be most beneficial, particularly isolated alterations just as well may destroy the "culture" of this industry and consequently be of quite disastrous effect.

4. In this contribution one of the main emphases was put on intellectual factors and their influence on (or determination of) industrial culture. This should not be interpreted as signifying that these are the most prominent and decisive aspects of industrial culture. However, the archetypical cognitive style of the "human actors" of individual industrial cultures is really of great importance and relevance in explaining differences in product development and especially concept design (which was, after all, the main objective of this study); furthermore it seems that this perspective did not receive its fair share of attention in the hitherto discussion of industrial culture and related studies.

5. This contribution almost prototypically illustrates one general problem of international comparative research which must be clearly denominated. In trying to depict and characterize certain features and phenomena of industrial cultures one has to negotiate a tightrope walk between really increasing mutual understanding by illuminating and explaining some culture-specific differences on the one hand and formulating generalizations which reduce the richness and diversity of a culture to normative statements and may even lead to prejudices or preconceived ideas on the other. However, this dilemma seems immanent to cross-cultural research, the only solution being awareness (of this problem), conscientiousness, and consequently exercising restraint in the formulation and distribution of judgements.

6. Even considering all the above-mentioned constraints, the concept of industrial culture will eventually prove superior to other approaches of international comparative research, as it attempts to take the full complexity of the interrelations and interactions between culture and industrial practice into account. And difficult as it may be, this is definitely not just the best but the only way to approach a more natural understanding and appreciation of the complexity and interrelatedness of "industry" as well as of "culture".

The above discussion summarizes and interprets some interesting differences in the product development styles of Japanese, American, and German machine tool companies.

The Japanese emphasize on the accumulation of information and constantly foster the ability to transfer this information quickly and efficiently into marketable machines. Their approach was therefore termed "synthetic".

The American designers tackle a problem instantly and with a concrete hands-on approach; they try to move on a direct way towards reaching the intended goal. Their proceeding was thus paraphrased as "pragmatic".

The Germans devote a considerable amount of time trying to understand and define the problem; they approach the goal comparatively slowly but thoroughly. This way was described as "analytic".

Though this report is based on the analysis of a comparatively small sample of firms, these characterizations can be traced like a central thread in different aspects of the product development procedure in the respective countries.
However, this paper must be regarded as a still "virgin" basis for discussion. At this point formulating catchy conclusions that transcend the mere description of product development styles is positively too early. I would thus be happy if people from the machine tool industry in particular would provide critical feedback and help improve on the preliminary implications of this paper. This is especially important in order to lay a solid foundation for a succeeding concrete discussion and formulation of practical strategies to improve on the existing situation.

A request for further cooperation, rather than some mildly enlightening phrases, therefore terminates this contribution.

References

1. Moritz, E.F.; Ein Vergleich von Strategien und Vorgehensweisen in der Produktinnovation in Japan und Deutschland am Beispiel des Werkzeugmaschinenbaus; Doctorate thesis, Intelligent Systems and Mechanical Engineering Laboratory, Tokyo Institute of Technology, 1994

2. Brödner, P., Schultetus, W.; Erfolgsfaktoren des japanischen Werkzeugmaschinenbaus; analysis of the RKW, the metal workers union, and the German machine tool builders association, Eschborn 1992

3. Moritz, E.F.; Die Konstruktionsausbildung in Japan im Vergleich mit Deutschland; Diploma thesis, Laboratory for design in mechanical engineering, TU München, 1988

4. Nomura, M.; Social conditions for CIM in Japan; International Conference on "Company Social Constitutions on the Pressure to Change", Berlin 1990

5. Tokunaga, S., Altmann, N., Nomura, M., Hiramoto, A.; Japanisches Personalmanagement - ein anderer Weg?; Campus, Frankfurt/New York 1991

6. Noble, D.F.; Forces of production - a social history of industrial automation; A. Knopf, New York 1984

7. Holland, M.; When the machine stopped; Cambridge University Press, Cambridge 1989

8. Moritz, E.F., Ito, Y.; Computer Aided Production Management - Innovation and Design; Sixth International Conference on Computer Aided Production Engineering, Edinburgh 1990

9. Moritz, E.F.; Konfuzius - Japan - Technik. Ein Alter Hut neu aufgesetzt; in: Wissenschaftliches Jahrbuch 1991, Deutsches Museum, München 1992

The Machine Tool Industry in Germany and the United States from the Perspective of Industrial Culture

by Gabrielle Laske

1 Introduction

The understanding of technology as based on and emerging from social, economic, and cultural conditions shows configurations of influencing aspects which throw light upon technological developments in a new way. The following results of a study on U.S. and German CNC machine tool design are a starting point for perceiving technological development and technical artefacts from an industrial cultural perspective. This article also points out how persistent and dynamic forces of an industrial culture effect implementation and design of technical artefacts. They either become an advantage or hindrance to competition on the global (CNC machine tool) market. Existing structures like the organization of workforce and production systems contain persistent elements. Performing people seldom notice the underlying values and norms in these structures and themselves. 'As nearly all our mental programs are affected by values, nearly all are affected by culture, and this is reflected by our behavior. The cultural component in all kinds of behavior is difficult to grasp for people who remain embedded in the same cultural enviroment; it takes a prolonged stay abroad and mixing with nationals there to recognize the numerous and often subtle differences in the way they and we behave, because that is how our society has programmed us.'(Hofstede, S. 23)

Dynamic forces put pressure on structures, institutions and their foundation, the traditional pattern and values. This pressure either occurs through social, economic or political events or as a combination of all. The Japanese challenge on the world market caused many changes in U.S. industry. Right now Europe might face a similar threat from the Japanese which will shake traditional behaviours in manufacturing and industry. Changes like the aging of workforce and demands for skill upgrading by increasingly implemented automation will have an input on machine design and applications. Engineers have to take these changes into account if their products are to be accepted in the long run. In general there is a poor awareness of the necessity to relate technology and the development of technology to a context of social and cultural conditions. The concept of industrial culture provides an opportunity to focus attention on this area. Therefore the following prospects are spotlights on industrial cultural determinants in the U.S. and Germany.

2 The Implementation of Advanced Technology in the United States and Germany

Industrial culture can be compared to a body or cultural medium in which technological innovations are implemented. This medium causes limitations and offers advantages by adapting technology to manufaturing and industry. Some specific American and German conditions one has to face while introducing the latest technologies shall be pointed out.

Traditionally the U.S. Air Force, aerospace and automotive industries are the major buyers of machine tools. They had an important impact on the American machine tool industry and the development of NC and CNC machine tools. Nevertheless, there are thousends of small and medium-sized machine shops for metal fabrication which are the major consumers of all machine tools (shops having from 1 to 99 employees accounted for 67% in 1989, source: Economic Handbook, NMTBA). Less than 11% of these metal fabrication facilities work with CNC (Source: Study of the U.S. Department of Commerce, 1989). Another study from The North American Manufacturing Technology in 1989 is concerned how implementation of advanced technology in U.S. plants does not even meet modest expectations. The expectations of the use of CNC machines, for example, were 50 to 52%. But: *"...the process of technology implementation in U.S. factories has been startlingly dissappointingly modest." ... "Relatively old and unsophisticated technologies characterize the U.S. technology base as limited and immature.'* (Manufacturing, NCMS, S. 307) One main reason for the creation of an atmosphere of avoidance and the lack of advanced technology in U.S. plants can be found in insufficient implementation. U.S. management scarcely analyzes scope and circumstances required to ensure successful implementation. They underestimate the need for restructured work organization, skill upgrading or training of their workforce to guarantee successful operations with new equipment. In general U.S. management has too little understanding and education about technology and its concerns.

In Germany the automotive, armament, and machine-building industries are the largest customers for CNC machine tools. Like in the U.S., small and medium sized machine shops have the largest share of consumption of CNC machine tools. The diffusion of advanced technology in German plants exists on a higher level than in the United States. The rate of the implementation of CNC machine tools is approx. 15%-20%. One expects an increase in automation at a ratio of 1:1 with already acquired technologies during the next few years. One third of these technologies will be a CNC machine tool. Manufacturing with advanced technologies is still considered to be at its starting point and is increasing at a slower pace than expected. A main need will be in the areas of metal fabrication and machine-building facilities, followed by the automotive and aerospace industries. German management is interested in automation and supported by a skilled engineering staff. In the past very ambitious automation projects like Computer Integrated Manufacturing were implemented and often proved to be more ambitious than effective and matured. Automation euphoria sometimes overweighs practical applicability.

3 Workforce and Skill requirements

At the beginning of this century F. Taylor developed his scientific management in the United States, where the U.S. industry consistently applied and practiced this organizational approach. The restraint of workers' intellectual capacities and a dissection of work processes into small units deprived man of the whole production process and eliminated the dignity of his work. U.S. companies still hire or fire their workforce when business rises or declines. Usually they rarely invest in training and further education. As a result, a shortage of skilled workers on a national level and also in many companies was created. Over many years these deficiencies become obvious, especially in applications of advanced technologies. While the U.S. labour market is not capable of providing industry with sufficient skilled workers, moreover an erosion of the workforce will affect the U.S. industry during the next few years. A dynamic change in its structure is taking place. Once the white male formed the majority of workers in industry. Studies show, 'that minorities, immigrants, and women will account for 80 percent of the future entrants into the labor force.' (Manufacturing, NCMS, S. 268)

Traditionally the U.S. worker is trained on the job. Work in industry is not a profession one needs to be educated for and the U.S. education system only rudimentarily prepares for jobs in industry. Operators who run a CNC machine tool have to show initiative to prove themselves as skilled workers, while working hard and seeking further skill upgrading opportunities if they wish some kind of career. U.S. society's traditional devaluation of work in industry is changing very sluggishly and still affects habits and actions of a major part of the U.S. population.

Now Taylor's scientific management is perceived as an obsolete concept. Many corporations are looking for new methods of workforce management. Teamwork is a key word and experiments in this area are numerous. Reorganization of production and organizational structures that emphasize teamwork lead to dismissals of middle management. A tendency that places programmer on the shop floor evokes the resistance of white-collar workers for they perceive these changes as down grading. The persistency of traditional status and structures are great hindrances in the process of reorganizing corporations and workforce.

The implementation of advanced technology alow leads to reconstruction of organizational and production concepts. But like manufacturing, technology is scarcely accompanied by US management's understanding of how to transfer and integrate it into their plants, new organizational concepts and workforce management usually face the same shortcoming. U.S. management also underestimates education and training, which are necessary to establish successful teamwork.

As already mentioned, the American use of technology is disappointingly modest and increases at a very slow pace. The lack of skilled workers is an obvious hindrance and will limit the automation of U.S. plants during the next few years. If U.S. enterprises wish to establish technological equipment on a higher level, they need to make large investments in education and training, for "the quality of American public education continues to plod along with antiquated systems, based on a model of society as it was over 100 years ago." (Manufacturing, NCMS, S.269).

The public education system does not teach for skills like programming machine tools or responsibilities for larger production units. Antiquated education systems and outdated production systems restrain the obvious necessity of operator training. Moreover, outdated production systems preserve the Taloristic production concept. Finally skill upgrading or deskilling of operators very much depend on management's philosophy and how it wants the factory to be run. There is probably a global trend to have operators programming CNC machine tools for small batch sizes. Because of the relatively old and unsophisticated technology in U.S. plants, a great deal of NC machine tools still seem to be used. M. Kelley mentioned "I ...found that, as of the mid-1970's, NC operators never programmed their machines. But this is not surprising, given the logistical problems and idle machine time that would be involved in operator programming of this pre-micro-processor version of technology" (M. Kelley, 1986, p. 227). In her study about skill upgrading or deskilling, M. Kelley showed that the workplace for operators has changed slightly since CNC machine tools were introduced. "The greater the NC share of all PA tools used in a plant, the more likely management is to depend on programming specialists or other white-collar workers in programming departments off the shop floor, rather than allowing operators to leave their machine stations to do programming. When all PA tools are of NC-type, the chances of blue-collar programming are therefore only 1:1.8. When NC-tools are half of all PA tools, there is an even chance (1:1) of blue-collar programming. When a plant has only CNC-tools (NC share=0), the chances of blue-collar programming are best at 1:1.8" (M. Kelly, 1990, p. 201).

NC was developed while scientific management was the major production concept in U.S. plants. NC technology was designed to state the dissection of white-collar intellectual work in the office and physical work done by blue-collar workers on the factory floor. CNC technology, on the other hand, offers various possibilities of workplace organization. But as long as NC machine tools run in U.S. factories, the Tayloristic method of production will continue to be practiced and operators' work will be on a low skill basis. Advanced automated technology demands skilled and intellectually involved workers. The current situation in U.S. plants shows no distinct tendency as to where workforce qualification and organization is going and whether operators will become more skill upgraded and responsible in the whole production process. This situation also reflects disorientation and incoherency of current American manufacturing.

In Germany the implementation of automated technology is also still at an early stage but on a higher level and with a different scope than in the United States. Lathe operators, milling machine operators or mechanics, for example, were clearly defined occupations requiring training for a period of 2-3 years. A reorganization of industrial occupations and their training system was initiated 5 years ago to keep up with the needs of modern production systems. German craftsmen in manufacturing gain social recognition. Although these occupations offer an approved status, there is a noticeable tendency towards academic careers. Successful young German craftsmen like to take the chance of further education in 'Fachhochschulen' or special university courses. This academic bias is one of the reasons why the German labour market has a shortage of qualified workers. The implementation of automated technology inevitably demands skill-upgrading even for relatively well trained-workers like in Germany. The German labour market cannot provide enough workers for these needs and the education system is struggling to keep up with the demand to train workers for the challenge of automated manufacturing.

In Germany corporations are also starting to experiment with new concepts of production and work processes. Despite the efforts to reorganize the education system, the traditional definition of manufacturing professions hampers this process and the idea of teamwork is being applied very slowly. Small and medium-sized German companies often are family owned. This structure preserves authoritarian and hierarchical attitudes which are deeply rooted in German culture and promote a Tayloristic organization of work. Professional status and authoritarian thinking are persistent forces which slow down the process of reorganizing work and production. They also obstruct the whole range of skills German workers could offer in teamwork.

Because of the shortage of skilled craftsmen, German firms tend to hire cheaper uneducated workers. But despite the employment of uneducated workers, the demand for skilled craftsmen for automated manufacturing remains acute. In the future, internal training programmes of corporations as well as new curriculum concepts from external sources will become more and more important in German industry.

The way in which the necessary modernization of outdated manufacturing equipment in East Germany will be carried out remains uncertain. An approach similar to that applied in the United States is possible, where initiatives try to implement advanced technology in simple and user-dedicated designs which do not require high financial and personnel investments. It will be interesting to see how long the East German industrial culture, based on its very different history, will remain a defined configuration. A study about this process will provide important insights how industrial culture works. During the next 10-20 years East Germany's industrial culture either will have merged with Western Germany's or will have maintained and developed an autonomous profile.

4 Engineers and design engineers

For many years the United States has faced a considerable shortage of engineers. Insufficient careers available for engineers have led to this shortage. U.S. entrepreneurs still regard engineering work as unimportant. Limited perspectives (over 5-7 years a talented engineer is able to reach the point of his career, where no further promotion is possible) make engineering careers unattractive. "The perceived lack of future opportunity in engineering and the relative appeal of finance, marketing, and general management career ladders, curtail many promising technical careers. And many of the best technical people are lost by promoting them into management positions" (Manufacturing, NCMS, p. 287). The low education level of American engineers has critical effects on the U.S. industry. "Too many of America's engineers are working with obsolete bodies of technical knowledge. For some time, a large portion of America's engineers have not had the level of skills required to do their jobs. ---* In the automobile industry, more than half of product and manufacturing engineers are operating with a knowledge base older than 10 years. ---* Some executives have discovered that orders they gave as many as five years ago to automate their production systems have not been carried out. In some cases, plant engineers did not know how to do it. In others, unwillingness to accept technological risk as a condition of technological advancement prevented modernizational efforts" (Manufacturing, NCMS, p. 288-289). The lack of engineering capacity in the United States is home-made. The American education system was reluctant to build up adequate engineering resources and U.S. corporations were not interested either. They preferred to send their engineers to 'short courses' and thought this to be sufficient to keep up with the knowledge about latest technology. Industry underestimated the need for qualified engineers. The result is a shortage of creative and committed engineers who introduce and speed up the application of advanced automation technology. Generally U.S. factories favour working with simple, easy-to-operate-machinery. They want it to be reliable, robust, low-priced and a guarantie of fast amortisation. Complex CNC machine tools or flexible manufacturing systems are often not run at their highest capacity because of a lack of qualified operators. If U.S. corporations want to implement automation technology on a high technical level, they have to intensively invest in education and training.

In Germany words like 'the art of German engineering' and 'the pride of German engineers' are well known. They refer to the high self-esteem and social status engineers enjoy in Germany. Engineering careers are attractive and many branches in industry and trade offer interesting promotions. They are highly paid and work under much more favorable conditions than their American colleagues. Earlier and more definitely than in the U.S., German industry realized the necessity of automated production systems. Qualified engineers will accompany the expected increase of automation during the next few years. The professional future for German engineers remains to be promising.

A socially highly valued engineering science will develop different technical products and standards than a science pushed into a marginalized position like in the U.S.. Therefore it is not surprising when German CNC machine tools are characterized in the United States as being overengineered. Therefore German CNC machine tools are only sold to niche markets but standardized German machines tools have no chance of succeeding in this area.

Demands of German machine tool users at the moment also appear to be changing. They are starting to reject overly ambitious technology. The 'German art of engineering' favours the bias to realize latest and high sophisticated technology in construction and design. As a consequence, German machine tools are very expensive and require highly qualified operators to run the machines. The US machine tool market has labelled these machines as overpriced and German customers are also beginning to ask for lower priced machines. Paradoxically German customers themselves have partly promoted the trend towards complex automation and specialized machines. Profound technological know-how and precise requirements make these customers sophisticated partners for machine tool builders. Traditionally the relationship between German machine tool users and producers has been very close. It turns out that in the long run the overemphasis on this type of relationship has become a disadvantage. A German engineer explained:

"German design engineers are used to listing to their customers. Actually this seems to be something positive. But - he is listening to everybody. This approach cannot lead to the development of simple, standardized machines."

5 The U.S. Machine Tool Industry - Profile and History

At the end of the 70s the U.S. machine tool industry lost its leadership on the world market. American insiders held that a complex of structural, economic, institutional and technological reasons was responsible for this. They do not exclusively relate to internal aspects of this branch. Aspects like insufficient education of engineers and operators and the lack of interest in research and development of basic technologies launched the process of decline in the U.S. machine tool industry. U.S. experts have recognized the following points as reasons which led to this downsizing.

Many U.S. machine tool companies were family owned at the beginning. They settled in areas where their main customers, the aerospace and automotive industries, had build plants. U.S. machine tool companies had to deal with cyclical vacillations in their market, as the machine tool industry all over the world does. In comparison with other industrial branches, the machine tool industry is a very small sector. The dependency on cyclical sales as well as its unimportance in relation to the total gross production made U.S. machine tool firms unattractive for financial investments. Lack of financial resources was a chronic situation. It turned out to be a hindrance for investment in research and development of new products although other reasons for this kind of neglect can be found. Even today a basic attitude in U.S. industry is the reluctance to understand that investment in research and innovation of manufacturing is necessary to keep up with world market quality and technological standards. Management in the U.S. machine tool industry was also guided by this attitude. With the strategy of backlogging they tried to balance cyclical vacillations of high or low orders. At times with a rush for machine tools, production was not fully adjusted to these demands. When orders decreased, they could still clear off remaining orders. This strategy made American machine tool customers wait for 1 to 2 years to receive their machines.

At the end of the 70s the U.S.government passed a law for emission regulation of automobiles. The U.S.automobile industry had to rebuild almost their whole production systems. At that time, the rush for machine tools increased tremendously. Machine tool factories proved to be very profitable and expanded their production on a large scale.

Between the 60s and 80s U.S. finance and business strategies were to make money by buying and selling companies. "In the late 1950s, they (U.S. companies) began diversifying, buying and selling companies in much the same way that investors trade stocks and bonds. As a result, management rarely invested the time and money required to fix troubled companies. When one of its operation lost market share or saw its margine decline, management simply sold it and used the proceeds to buy another company" (Manufacturing, NCMS, p. X).

While U.S. machine tool companies seemed to be profitable, conglomerates bought them but were only interested in quarterly profits. Investment and innovation in R&D or manufacturing facilities were neglected. During the recession in the early 80s most orders for machine tools were cancelled. All of a sudden, machine tool companies had to dismiss hundreds of workers. These companies were no longer profitable, conglomerates sold them and left them without any financial resources, outdated manufacturing hardware, and product lines which could not meet quality and standards on the international machine tool market. For a long period of time they lost the ability to react to changes on the domestic and international market. After this decline only few survivors remained in U.S. machine tool business.

Not only the above mentioned business strategies restrained innovation and progress of U.S. machine tool builders. Many of them never sold their machines on foreign markets and so never encountered the challenge of foreign competition because the U.S. market is huge and seems to provide sufficient business. Up to the beginning of the 90s, a tendency to limit oneself to domestic markets can be observed. A member of the National Association for Manufacturing Technology (NMTBA) noticed:

"We have many members who never had export. They just think in terms of the United States."

Financial, organizational, and technological neglects had weakened the U.S. machine tool industry to a point where it became easy for Japanese machine tool builders to enter the U.S. market. Moreover, the Japanese offered service to a degree unknown to American machine tool users. The following experience of a NMTBA member is cited as an example.

"I was in a shop up in the country in New York. This guy had two machines in a barn and he had an American built machine. He had a lot of trouble with it. Had trouble getting the service man to come up in the country. He bought a Japanese machine and said - they would fly a man from San Francisco over night. He would rent a car in Albany in New York and he would drive up and he did have the guy here the next day. He said: when he bought his second machine, he didn't even inflight an American to bid on it."

Now Japanese CNC machine tools dominate on the world market and have the largest market share in the United States. The former leader of machine tools on the world market remained on a very diminished scale. Many surviving U.S. machine tool builders choose to emphasize products for niche markets like complex machining centers, high volume or high precision machines. U.S. firms which try to compete on the low price standard machine tool market meet intense pressure from foreign competitors.

From the perspective of industrial culture traditions, social norms, and patterns can be pointed out which troubled and partly destroyed the U.S. machine tool industry.

- a national self-sufficency and lack of interest in the global machine tool market
- ignorance of the necessity to invest in research and development for basic technologies
- lack of investment in new manufacturing technology. Obtaining quarterly profits is of importance. The regeneration of production resources is of no use
- underestimation of the importance of qualified workers

To sum up, these points show how U.S. management failed to develop time and global perspectives. Concepts of innovation, research and development and long-term financial commitment can only be built up in the context of long-term prospects. Lack of global perspectives resulted from the fallacy that the domestic market would provide sufficient customers for American machine tools; export was of no interest. But the huge U.S. market was open to foreign products for all important markets are exposed to global market conditions.

6 The German Machine Tool Industry - Profile and History

Most German machine tool companies were founded by technicians or engineers and, in some cases, are still family owned. German machine tool builders traditionally have close relationships to their customers. The strength of German engineering science combined with highly qualified craftsmen promote the implementation of automated technology and the development of sophisticated technology in the machine tool industry.

After World War II major parts of German manufacturing systems were destroyed or dismantled for reparation payments. During the time of economic regeneration German factories were able to build up modern production systems. The relapse to old pattern production and machinery was nearly impossible. Machine tools construction designs also were taken for reparation payments and new machines had to be designed and constructed. In the 70s the rebuilding of the German machine tool industry reached a point where it became well-known for production of high quality and highly advanced CNC machine tools.

In the 80s the German machine tool industry was very successful in exporting mainly to European countries. A declining business cycle in 1991/92 followed by little interest on the part of customers in investment for CNC-machine tools made weaknesses and deficits in this industry stand out which have been latent for years.

The German machine tool industry has about 3 different approaches towards technology and marketing. 1) specialized machine tools offering high volume, high precision, 2) machine tools with high technical performance, sort of semistandardized machine tools because they can be extensively customized, 3) standardized machine tools. Corporations with specialized machine tools are now more successful than corporations who build standardized or semistandardized machines. The current sales crisis is partly home-made but only few of the machine tool builders are sensitive to this fact. The interim results of this study show that German machine tool industry is facing the dilemma of being very strong in customized and specialized machines but that it has neglected the high volume market of standardized CNC-machine tools. This market increasingly seems to be lost and German machine tool builders self-critically describe their machines as technically too complex.

Also German engineers need much more time than Japanese do to develop and construct new CNC machine tool products. In many companies the order to develop a new product is given to the design engineer or the design engineers' office. This order is very often not based on precise marketing analysis. In some cases in the past reason to create a new product was plainly because of the next fair, where one wished to present a new product. Germans have the tendency to develop machine tools not so much as optimum and low-priced production tools, but as some kind of technical works of art. This approach towards technology development is primarily interested in product's technological aspects and much less in the practical and enviromental side of its application. It is expensive and time-consuming because users' criticisms and corrections are then received at a late development phase of the first prototype. Changes at this stage increase prices of machines more than at an earlier stage.

The U.S. machine tool market perceives German machines as too complicated, with too many technical options one does not need but has to pay for. The assessment about expensive and technical oversophisticated standardized machines also comes as feedback from other machine tool markets. Even in Germany, and particularly because of the recession, machine tool users require a different kind of product. The actual trend favors standardized machine tools with lower complexity, low price, high quality and user-dedicated controls.

During the last few months efforts have been made to design simpler, low-priced and high-quality machines. This is true even with machine tool builders of specialized machines. But there is almost no discussion about high control expenses, but rather expenses of software development.

7 The Relationship betwenn U.S. Machine Tool Builder and Control Manufacturers

For many years the nation where numerical control was invented failed to carry on this development for low-priced and user-dedicated adaptation. Research and development of the first numerical control was financed by the U.S. Air Force. After Parson, the inventor of this technology, withdrew from the project, the Servo Mechanical Laboratory of the Massachusett Institute of Technology (MIT) received almost unlimited financial support to finish this development. Finally the laboratory had developed a technological, highly complicated solution of automated machine tool control. Demands such as low price and easy operation were unimportant. The implementation of NC technology by U.S. Air Force suppliers was financed by the U.S. Air Force. Orders were only placed on condition that NC was implemented. NC history demonstrates a development process that was not related to commercial use or demands of the machine tool market. An engineering approach took place that was not interested in the needs of U.S. manufacturers with their specific production and work organization conditions. Simple and practical use of this technology was not taken into consideration. This kind of technological development was only possible because of immense financial support. These features were reflected in commercially produced American controls. Complicated technology, computers with great hardware expenditure and, because of high complexity, low reliability in production were characteristic of American controls. "These problems of complexity and reliability carried over into the CNC era. The two major domestic manufacturers of controls, GE and Allen-Bradley, were not especially responsive to either the builders or end users, and kept building very sophisticated controls; indeed, one insider described GE as seeking to make the "ultimate control", which kept different and non-communicating parts of the organization"fooling around" with exotic materials, lots of board layers, and highly complex software". (MIT, Working Papers, p. 25)

With the implementation of CNC, NC technology became available for broader applicability. Expansive computer hardware and programmers became less important in the use of CNC machine tools. But U.S. manufacturers of control systems persisted on their outdated design even when CNC technology offered cheaper and simpler solutions. "GE remained committed to hardwire controls well after microprocessors became available in the middle 1970s. It did not have a solid state CNC control on the market until the end of the decade. ---...if domestic builders did not choose to put up with hardwire controls or excessively sophisticated controls that matched their machines, they were forced to go overseas or make their own; - "(MIT, Working Papers, p. 25)

Allen-Bradley favoured a concept of intense customization that increased cost and diminished reliability. This trend made U.S. machine tools very sensitive, especially with regard to their controls. U.S. machine tool builders tried to develop and produce their own controls or bought small control manufacturers for in-house efforts to eliminate this weakness. A U.S. machine tool engineer described the former situation: "The electronic and controls that we had on the American machines were not reliable, they were not as reliable as Japanese. The Japanese controls were just comming into the marketplace. They were more reliable and when our customer had a machine that wasn't running, he did not care what control was on. What he was saying was: the Warner and Swasey or the Cincinnati Milacron - that machine is not running. But we have many American controls on those machines. Other customers who bought a Japanese machine were telling us: You've got to put a Japanese control on your machine, because these machines are so reliable, they run forever and never stop. So we lost a technical edge here. Not necessarily the machine tool companies, but the machine tool companies were a victim of the electronics industry that was not producing reliable equipment."

U.S. control manufacturers extensively lost market shares to the Japanese. FANUC now holds 50% of the market for controls. The efforts of American machine tool builders to build their own control in general were not very successful and were also very expensive. Therefore most of these in-house efforts failed and U.S. machine tool builders were forced to use foreign, more solid controls like Siemens and FANUC.

As a result of the American control evolution, many U.S. machine tool builders are selling their machines with several control options. The lack of a satisfactory U.S. control solution drove costumers to meet their needs on the international market and they expect U.S. machine tool builders to provide their machines with controls of their choice. Offering several options is a very customer-oriented approach and very costly.

American CNC contol history is not complete at this point. The Next Generation Controller Program (NGC), initiated by major U.S.-companies is aimed at developing a low-priced and open control concept with exchangeable hardware and software modules which are to become independent of self-contained control systems like FANUC or Siemens. The target to develop cheaper controls than FANUC led to considerations that to be lower-priced than the largest control manufacturer, one has to procure electronic and hardware components from the electronics industry. Controls made with commercially available components are cheaper than any control manufacturer with in-house production could deliver. The U.S. electronics industry is very strong. This strength was seen to be fundamental for the NGC concept. The preliminary outcome is a Next Generation Control composed of electronic parts one can buy on the market. Efforts were made to develop strong and compatible drives which ensure high performance. At present it is not foreseeable whether the Next Generation Control will revitalize the U.S. machine tool industry and leads to a regaining of international market shares.

8 The Relationship Between German Machine Tool Builders and Control Manfacturers

The German machine tool industry receives far more scientific support for R&D concerning machine tool construction and controls than the U.S. machine tool industry does. Frauenhofer Institutes in Aachen, Stuttgart and Berlin are well-known institutions involved in work on innovations for machine tool technologies. The scientific leaders of these institutes were committed to the NC mainstream tradition which came from the United States and was modified by the German tradition of highly qualified engineers and craftsmen.

The relationship between German control manufacturers and machine tool builders implies an unreflected contradiction between NC technology, based on the Tayloristic production philosophy, and German craftsmanship traditions, with their high level of qualifications and socially acknowledged status.

For many years, German machine tool builders and control manufacturers developed close relationships. Long-term cooperation with, for example, Siemens, Bosch, Grundig, Heidenhain, or Philips are common with most machine tool companies. Intimate knowledge about capacities, needs, advantages and disadvantages of each side was obtained. These partnerships are strong, solid teams. Controls developed under these circumstances respond very specifically to customer demands and are high-priced because of high development and production costs. Experts estimate these costs at about 15%-30% of the total machine tool.

Annually, Japanese control manufacturer produce 60.000 controls while Germans only produce 15.000. This is a considerable competitive price advantage for Japanese machine tools. Moreover, the Japanese lead in controls enables them to build lower priced universal machine tools even in the area of special and high-precision machines.

In the relationship between German control manufacturers and machine tool builders the bias towards special, sophisticated solutions continues to be favored. This results from R&D approaches which do not perceive technological artefacts as embedded in a social environment with specific production philosophies, workforce qualification and work organization. Rather, machine tools are seen and designed more as a brilliant work of art where electronic and mechanical engineers demonstrate their superior knowledge. This is doubtlessly a solid resource to develop high-tech solutions but at the price of losing extensive market shares. Siemens, representative for high end controls, and FANUC, which stands for high volume, tried to cooperate in a joint venture in the 80s. The cooperation was not successful. A German software engineer reported:

"The control, developd by this joint venture, was not as successful as both control manufacturers had expected. Probably their mentalities also did not fit. They have far too different concepts and ways of thinking. Each control approach differs considerably. FANUC emphasizes volume, guarantees reliability, but special technical features rank second or third. FANUC controls are high-volume products. Siemens wants to provide controls to a wide variety of customers and therefore tries to respond to as many demands as possible. Often this goes wrong. FANUC chooses a single concept and then markets it. Nobody is bothered when certain feature are not available. If Siemens acted like this, every German customer would be annoyed. Customers expect Siemens to respond due to its corporate philosophy. With FANUC controls they don't have this kind of expectation."

A high social status of craftsmanship as well as union's commitment for this profession, education and qualification system shaped German concepts for control and programming in a unique way. Discussions and efforts to develop on-line controls which increase the use of craftsmen's skills and know-how could not have taken place in the United States. Many German machine tool builders try to optimize CNC machine tool operators' qualifications and work by applying appropriate hardware and control design to their machines. They also try to adapt special technical features which satisfy engineers' ambitions. These are contradictory tendencies which often lead to failure. The approach of Siemens control 800 mainly emphasized high technology. A German software engineer's comment:

"There is a good description one of our craftsmen gave to Siemens control 800: You can fly to the moon with this control, but it is not suitable for shop floor programming. This control can drive very complex production systems, but is absolutely insufficient in use for simple production processes."

The close relationship between German machine tool builders and control manufactures has created considerable knowledge in the development of software and programs. Here is the link where craftsmen's skills and the advanced science of engineering come together if they are able to connect and to communicate. Unfortunately this advantage does not pay off. To furnish machine tools with this kind of semi-in-house controls can only be done in small batch sizes and remains too costly.

9 CNC Machine Tool Builders and New Technologies in the Light of Industrial Cultural Determinants

"I believe that norms change rarely by direct adoption of outside values, bur rather through a shift in ecological conditions: technological, economical, and hygienic. In general the norm shifts will be gradual unless outside influences are particularly violent."(G. Hofstede, p. 26)

Industrial cultural determinants described and pointed out in discussions with U.S. machine tool builders show values, norms and patterns which are opposite to constructive implementations of advanced technology. Basic values and norms shape the ideas and action of many professions, for example, managers, operators and so on. Failure to innovate production systems and neglect of workforce training result from social disdain of those U.S. industries that emphasize basic technologies. Awareness that successful and modern production is fundamental for a high national standard of living was not able to spread among the broad American public. Acknowledged technologies have to be ambitious and need to aim at least to make the first step on the moon. Technological evolution means to conquer the unknown but not to improve already applied technologies. Values and norms are persistent forces which try to maintain an already existing status. But they are also constantly effected by internal and external changes. The overwhelming Japanese presence in the area of basic technologies on the U.S. market puts strong pressure on existing industrial attitudes. A decline of living standard, experienced by large portions of the U.S. population, causes reactions and reflections towards traditionally repeated patterns in production and development. In order to survive, U.S. corporations have to seek new forms of work organization and ways to train their workforce. A shift in values and norms can be observed that does not necessarily eliminate traditional values, but rather modifies them for social and economic changes. In search of a new concept the U.S. machine tool industry has initiated an ambitious project to develop a third control generation. It was imporant to receive impulses from the strong U.S. electronics industry.
Unlike the MIT approach to develop numerical control unrelated with market demands, this time a concept was chosen that focused on low price and open structures to make sure that future software and hardware developments can be integrated. Customers' low budgets and the implementation of further development phases were taken into account. Considering the above described industrial situation in the United States, an appropriate American control system should be rugged,

reliable, and easy to program to meet the low education of the future American workforce. Especially in small and medium-sized shops, these features will be necessary to make CNC machine tools an acceptable and useful instrument. For many years U.S. society's and U.S. industry's emphasis on and overestimation of high technology and high volume production have led to a prevailing unproductive manufacturing base. But, in general, U.S. metal fabrication needs to concentrate on implementation and use of solid, easy-to-operate basic technologies.

Values and norms said to affect German industrial culture are:
1) a pride in quality and a desire to meet every demand,
2) society's assurance of the necessity of training and education,
3) traditional authority through rank, status and qualifications.
In order to produce high-quality products, skilled workers and adequate production environments are needed. The above mentioned German values support the adoption of new technologies. In general reorganization of work follows implementation of new technologies. Through work with computers, employees gain more insight into planning and organization. The restructuring of work and the making of production processes more visible conflict with traditional authoritarian structures. Middle management becomes cut down and craftsmen have to deal with increasing responsibility and higher demands for education and training.

German machine tool builders like to present their CNC machine tools as some kind of masterpiece. Every major German machine tool company has an especially styled showroom to present their machines to customers and visitors. During visits of several, even large, American machine tool companies not one single showroom for the latest CNC machine tools was seen. The showroom symbolizes German machine tool builders' and design engineers' pride and satisfaction to create advanced high-tech machines. Machine tool users with lower education levels feel overtaxed or regard certain machine tool features as unnecessary. Many countries do not have well-educated workforces like Germany does. Their operators and facilities are not able to make use of German machine tools. They will not buy German machine tools, as studies show, and even German craftsmen are under stress while working with expensive, highly complex machine tools.

Upheavals in the German machine tool industry are becoming visible and the engineering philosophy of only creating ambitious and advanced machine tool concepts is starting to become a hindrance for the whole industry. To follow this track means to lose extensive market shares. The global machine tool market does not mainly require complex high-tech but low-priced, easy-to-operate machines. German engineers once responsible for the success of the German machine tool industry have now uncoupled this industry from basic technological standards and shifted it into the high-tech domain.

For the future education of engineers an emphasis on engineering concepts that relate technological developments to cultural, social and enviromental contexts might be necessary. New production and organization processes need to be taken into

consideration.
The strengths of the German industry are highly qualified craftsmen, engineers, and progressive engineering science. A way to make a competitive advantage out of this resource has not yet been worked out.

References

Asdonk,Jupp/Bredeweg,Udo/Kovol, Uli
Die mikropolitische Arena technischer Innovation
Eine Untersuchung zur Technikgenese im Bereich der Produktionstechnik; Forschungsbericht, Bielefeld, 1992

Bailey, Thomas
Change in the Nature and Structure of Work: Skillrequirements and Skill Formation
University of California, Berkeley, 1990

Bendix, Ulf/Knuth, Jutta/Kuhn, Michael/u.a.
- Industrielle Metallberufe - Berufsreport 1A,1B,2
Bremen, 1991

Böhle, Fritz/Mielkau, Brigitte
Vom Handrad zum Bildschirm
Frankfurt, 1988

Brödner, Peter
Fabrik 2000 Alternative Entwickungspfade in der Zukunft der Fabrik
Berlin, 1985

Competing in World-Class Manufacturing
National Center for Manufacturing Science
Homewood, Il 1990

Das Erleben der Arbeit am Arbeitsplatz mit neuer Technik
Am Beispiel von CNC-Werkzeugmaschinen in der industriellen Fertigung
Projektbericht, Universität Hamburg, 1988

Hofstede, Geert
Culture's Consequences
Sage, Newbury Park, 1990

Kelley, Maryellen R.
New Process Technology, Job Design, and Work Organization: A Contingency Model
American Social Review, Vol. 55; 1990

Kelley, Maryellen R.
Programmable Automation and the Skill Question: a Reinterpretation of the Cross-National Evidence
Human System Management 6; 1986

Methods to Transfer Federal Technology and Improve Industrial Competitiveness
Executive Summary
Georgia Institute of Technology, 1991

Paschen,Herbert/Gresser, Klaus/Conrad, Felix
Technology Assessment: Technologiefolgenabschätzung
Frankfurt/New York, 1978

Rauner, Felix/Ruth, Klaus
Perspectives of Research in 'Industrial Culture'
Elsevier Science Publisher B.V., 1990

The U.S. Machine Tool Industry and its Foreign competitors
Working Paper; MIT Commission on Industrial Productivity
Cambridge, 1989

Industrial Cultures and Machine Tool Industries: Competitiveness and Innovation Trajectories

by Klaus Ruth

1 Introduction

The crisis in the world machine tool production industry, which has lasted longer than expected and is characterized by a production decline over several years that amounted to approx. 8.3% in 1991, with production revenue dropping to \$41.5 billion, and the still weak order situation in the German machine tool industry in particular are cause for a heated discussion over the situation and prospects of the German machine tool industry. The debate is roughly split into two camps, pessimists and sceptical optimists. The pessimists, particularly the group of "panic-mongers", refer to the Japanese machine tool industry, its unparalleled production success during the past two decades and a change taking place in the demand structure (in Germany) towards simpler, less complex CNC machine tools, the demand for which is currently being met with growing success by Japanese manufacturers. An indicator of this creeping development is the import/export ratio. The ratio between machine tool imports and exports has continuously worsened for the German machine tool industry since 1984: while the import-export ratio in 1984 amounted to 28%, it worsened to 46% in 1990 with a continuing trend in this direction (VDMA 1991). Thus, although the overall balance is still far from being negative, as has been typical of the U.S. for years now, a weakening of the internal production structure of Germany's machine tool industry appears to be taking place.

The sceptical optimists, on the other hand, see the current crisis primarily as another round of the up and down cycle typical of the sector and they do not assume that the present slack in the industry's activities represents a basic downswing into a large-scale decline in the German machine tool industry. They base their assessment, first of all, on the fact that the entire economy is in a (temporary) downswing, with the machine tool sector being no exception. Its strongest argument, however, rests on the presumed great innovative capacity of the German machine tool industry and its efficient production structure.

The findings presented below are intended not only to contribute to providing an evaluation of the innovative capacity and the special features of the production structures for the German machine tool industry, but also to assess the two leading machine tool producing nations worldwide through a comparison with Japanese findings so as to determine the prospects of both countries on the basis of their respective innovation and production structures. In addition to presenting general explanations, I will look at the essential results of a study of manufacturers conducted within the framework of the project, "Industrial-Cultural Determinants of strongest argument, however, rests on the presumed great innovative capacity of the German machine tool industry and its efficient production structure.

The findings presented below are intended not only to contribute to providing an evaluation of the innovative capacity and the special features of the production structures for the German machine tool industry, but also to assess the two leading machine tool producing nations worldwide through a comparison with Japanese findings so as to determine the prospects of both countries on the basis of their respective innovation and production structures. In addition to presenting general explanations, I will look at the essential results of a study of manufacturers conducted within the framework of the project, "Industrial-Cultural Determinants of CNC Machine Tool Production", in two countries, Japan and Germany. However the gentle reader might get the impression that as a matter of fact in this contribution more questions are raised than answered, and that is why this article also can be perceived as instructive for future research in the field.

2 Structure and Significance of the Machine Tool Industry

The machine tool sector comprises enterprises that manufacture products in goods groups 736.1 and 736.2 (in accordance with SITC II, the Standard International Trade Classification, Revision II of the UN). Goods group 736.1 includes mechanical engineering products made by means of metal cutting while classification 736.2 covers non-cutting forming work. Furthermore, manufacturers of mechanical and electronic components for machine tools are usually considered to be part of the machine tool industry. Since the empirical data presented below refer to metal-cutting forming work, i.e. to lathes, grinding and milling machines, the emphasis at this point will be placed on the structure and significance of metal-cutting machine tools.[1]

Despite its small size, an especially strategic importance is attached to the machine tool industry in all national economies in which such a sector has developed at all, in particular the metal processing industries. Machine tools for metal-cutting (as well as shaping) work are involved in the manufacture of nearly all products in the metal industry - either directly, via production of end products and semifinished articles on machine tools, or through production of other machines used for the manufacture of end products by means of machine tools. It is plain that the efficiency and competitiveness of manufacturers of end products as well as the quality and precision of the products themselves depend to a high degree on the quality of machine tools. The latter thus represent the production heart of manufacturing industries. They have the important function of being national developers and suppliers of technology; that is why it is eminently important for national economies to have efficient machine tool industries. The

1 The selected types of machines correspond to SITC categories 736.11-5 and 736.17-9 or, in accordance with the German systematic list of goods for production statistics, to goods groups 32112 to 5 and 32117 and 32118.

smallness of the sector is inversely proportional to its economic and innovative significance.

In 1991 the production volume for metal-cutting and shaping machine tools (excluding parts and accessories) had a value of approx. DM 15.156 Billion (incl. estimated share of new federal states in East Germany) in Germany and DM 19.299 Billion in Japan (*American Machinist* according to *VDI-Nachrichten* 13, 1992). The German value of production thus declined slightly in comparison with the previous year, a trend which will, in all likelihood, worsen significantly in 1992. This development does not appear to affect the Japanese machine tool industry to a comparable extent. While the German machine tool industry evidently undergoes sharp cyclical ups and downs (see Fig. 1 with the development of order entries), there is a smoother curve in the case of the Japanese machine tool sector.

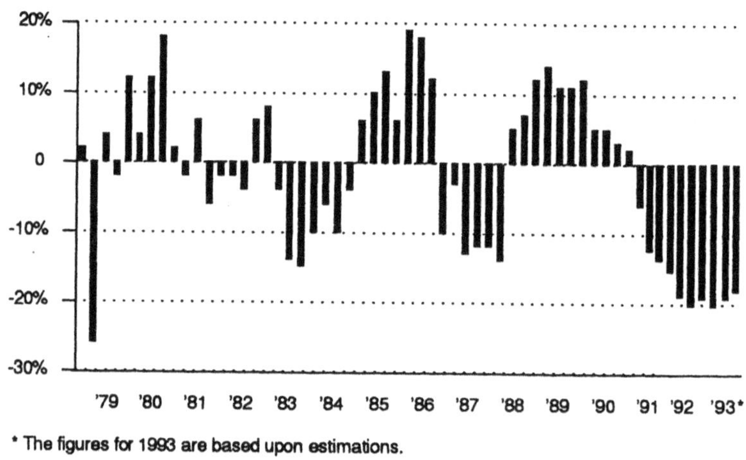

Figure 1: Order entry in the German Machine Tool Industry

The German machine tool industry employed roughly 102,000 people in approx. 400 enterprises in 1990. No other branch of the mechanical engineering sector has more employees (in 1990 a total of 1.16 million people were working in 6271 enterprises). The situation in the Japanese machine tool industry is somewhat different.[2] There are 393 enterprises (with more than 20 employees) that employ a total of 44369 workers operating in the particularly interesting area of metal-cutting machine tools. Because of the pronounced supplier structures, these firms are surrounded by a ring of innumerable small and tiny enterprises (with less than 20 employees) that produce components, modules and accessories for the machine tool manufacturers. The first impression that the Japanese machine tool industry achieves an approximately equal volume of production of metal-cutting machine tools with less than half as many employees is, at least to this degree, not accurate.

2 It must be emphasized that the German and Japanese characteristic data used here are only comparable to a limited extent.

A comparative look at the company size and employee structure reveals several special features[3]: the number of large companies having more than 1000 employees is significantly higher in Japan than in Germany; approx. 40% of the machine tool enterprises fall into the category of 100-499 employees while the figure for Japan is over 55% (at least in the sample presented here). The numerical strength of small enterprises in the German machine tool industry is also striking (see Figure 2).

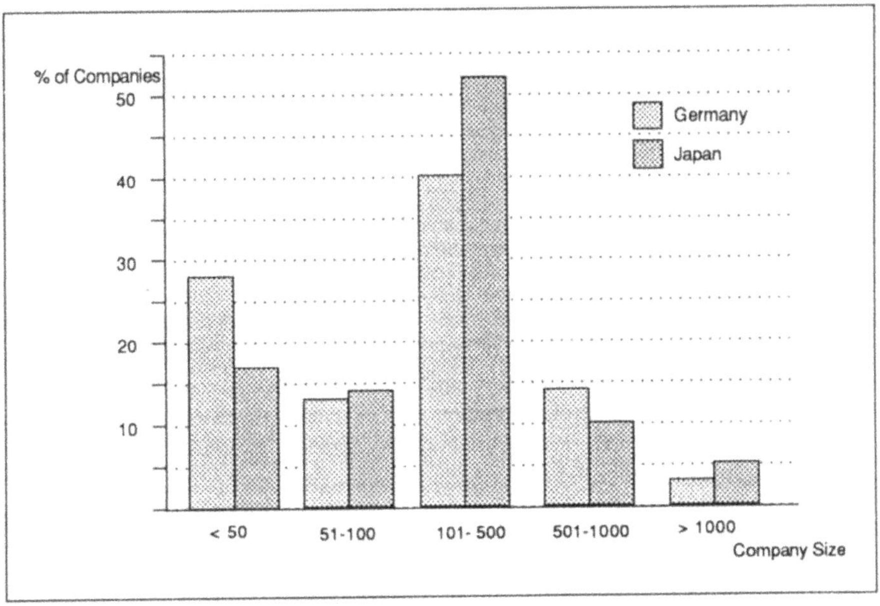

Figure 2: Structure of Machine Tool Sectors
(Sources: VDMA 1991 and JMTBA 1991)

In summary, the figures indicate a predominance of small and medium-size enterprises in the structure of the German machine tool industry. This hypothesis is supported by an analysis of the sales of the largest companies in the respective sector (see Figure 3). Figure 3 shows the three Japanese and German machine tool companies with the highest sales. The marked difference catches one's eye immediately.

The annual sales of Japanese companies in 1990 exceed those of Germany's leading companies in sales many times over; the cumulated sales of the three largest German corporations exceed the sales of the largest Japanese enterprise by less than 10%! If one continued down the ranking of company sales, other Japanese companies would slip into the list between the top three Japanese corporations and the German company with the strongest sales so that the first German company would be number eight.

3 It must be taken into account in this comparison that the data on the German machine tool industry take into consideration almost all relevant units while the Japanese figures cover only a fraction of the basic whole. The question of representativeness cannot be answered.

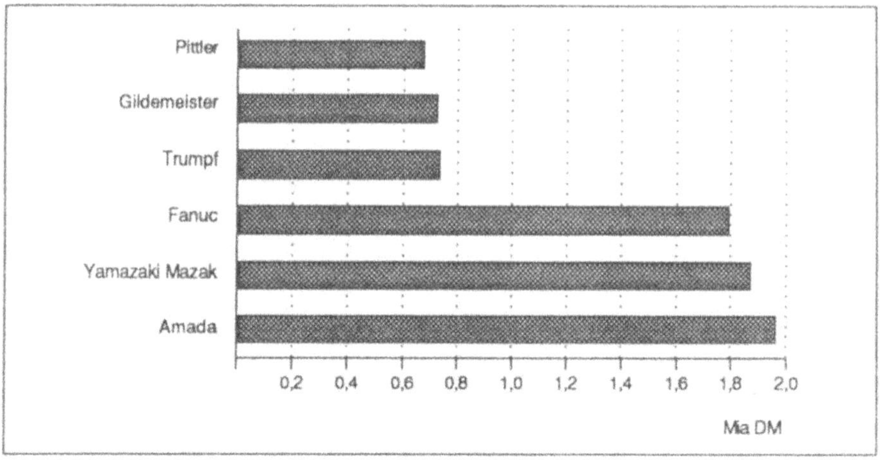

Figure 3: **Annual Turnover of selected companies by rank**
(Sources: American Machinist 1991 and VDMA 1991)

These obvious differences do not yet enable one to come to a definite conclusion as to the superior innovative force of the larger Japanese companies with higher sales volumes because it is always the small and medium-size enterprises that push forward innovations more quickly and flexibly than bureaucratized large corporations. However, international competitiveness, presence on foreign markets will undoubtedly be easier for companies with great amounts of capital and high sales than for smaller ones with low sales volumes. Although the above outlined data represent important structural features, they have little meaningfulness for the specific question regarding the prospects of the German and Japanese machine tool industry. Therefore, it will be attempted in the following to assess the prospects on the basis of selected results of the international CAPIRN machine tool producer surveys by examining more closely the actual innovative capability, innovation make-up and sources as well as the innovation processes themselves.

3 Innovative Capabilities, Innovation Processes and Typical Innovation Make-Up

The percentage of companies (of a sector) that implement product and/or process innovations would be a suitable general measure having limited significance. The average figure for all sectors of the processing industry in Germany for 1987 was approx. 75% (the proportion of innovators in 1979, on the other hand, was only 60%). A differentiation according to sectors shows an innovator share of 78% (1990) for the capital goods industry, in which the machine tool sector occupies a prominent position: the figures for machine tool production are over 80% (*VDI Nachrichten* 37, 1991).

Considering the composition of companies' **product mix** as an indicator for innovative capabilities we state that the dominating pattern of innovative companies usually operating in growing markets features a share of products in the stage of market introduction which outweighs those in the decline phase.

It has not yet been resolved whether the CNC machine tool production sector is a growing or stagnating market. Indeed the *Strategic Study on EC Machine Tool Sector* (1990) estimated an annual growth of the world market for machine tools of 4% p.a. by 1995 in real terms but reality reveals sluggishness and recession. If the machine tool builders can continue to develop an innovative product mix; i.e. the sales achieved with products in the market introduction phase overcompensate for the sales derived from "discontinued" products in the decline phase remains in the dark. This innovative characteristic of product life cycles can be regarded as the basis of their competitiveness. However, we must assume predominantly **incremental innovation processes**, both for the mechanical engineering field as well as in the case of the machine tool industry, i.e. innovative products are usually not new ones, but improvements and adjustments of existing products.

In connection with innovation and competitiveness, the question of the sources of innovation is of special importance - that is, the question of where triggers for certain innovations come from. Our original hypothesis formulated in the comparative project on industrial-cultural determinants of CNC machine tool production presupposed that **the relationship between CNC machine tool producers and customers represented an important source of innovation for the German machine tool industry**. We perceived traditionally close formal and informal manufacturer/customer relations as a peculiarity of the German industrial culture. These relations make tailor-made system solutions according to customer requirements feasible and in the medium-term these experiences are "productively assimilated" into (incremental) technological developments - so much for our hypothesis.

In the following the results of the standardized quantitative survey conducted among manufacturers of CNC machine tools will be made use of for basic verification of the industrial culture hypotheses.[4] The results presented thus represent initial indications that confirm the "suspicion at first glance" or lend plausibility to the working hypothesis. A concluding verification is only possible on the basis of the results of all substudies (i.e. manufacturer questionnaire, user questionnaire, interviews and case studies planned in the project). Nevertheless, it is meaningful to disclose initial results because important information can be derived from them for the substudies that follow.

In the further course of this report I will, whenever possible, compare the findings of the German study with results of the Japanese sample.[5] This

4 The sample comprises 29 companies from the machine tool industry that manufacture either lathes, grinding machines or milling machines. This segment of the industry is composed of a basic total of approx. 70 companies, which were written to in a postal survey. Since the sample taken shows a bias towards large companies, it was weighted accordingly. All frequency tables and the like reflect the results of the weighted sample - with the exception of the calculation of correlation coefficients: here the unweighted sample was the basis.

5 The comparison must be done very carefully and the results must not be overestimated, because comparing two samples which are not sure to be representative has to be conducted carefully. But anyway the data give hints and probably show important

corresponds to the comparative approach of the industrial culture concept, i.e. comparing and contrasting various technical and other characteristics and thus revealing differences that are most likely specific to the industrial cultures.

Design sources and influences

In the view of German CNC machine manufacturers, customer companies have the opportunity of influencing the composition of machine systems. Approximately 60% of the responses stated that it was possible to buy individually configurated machine modules (such as certain CNC controllers, a certain drive unit, certain tool handlers, etc.). This strategy of module combination seems to be oriented towards more standardized machines. Another 60% indicated the common practice of using requirement profiles (lists of requirements) as a means of influencing the shape and composition of the machine tool. The lists of requirements are primarily used for **adjustment design** and **design of variants**, but only in a small number of cases for a complete redesign, i.e. nonstandardized machine tools (INT01190A)[6]. About 12% of the companies replied that they utilize other paths of customer influence: particularly the special purpose machine tool producers (in our sample producers of milling machines) offer very extensive opportunities of customer participation, such as **joint conceptual design** with the machine planning department of the customer[7]; and in the case of control unit producers it was possible to shape the human/machine interface according to the customers' needs[8].

With regard to customer influence on machine system composition, the sample confirms our assumption that this is possible and is also made use of by the customers. In addition to offering and selling machines purely "of the shelf", manufacturers have increasingly gone over to "modularizing" their range of machines and presenting it to the customer in combination with individual, problem-based consulting services. This procedure allows machine tool producers, to a limited extent, to take advantage of scale economies with a high degree of specialization and flexibility at the same time! The customers benefit from this strategy by tailoring machine systems to their manufacturing problems - within certain limits but without the additional costs of redesign or adjustment design.

When comparing the findings in the German case with those in the Japanese sample, we find similarities regarding the significance of **requirement profiles**. In the Japanese sample 60% of the responding companies indicated lists of requirements as a means of influencing the design of machine tools. The category "others" shows approx. 8% but this figure defies interpretation because these other means were not specified. In the market segment comprising standardized CNC machine tools the tendency to combine machine modules is not as widespread as in

tendencies. The Japanese sample comprises of 45 responding machine tool building companies.

6 INT01190A and similar indications in the following refer to interviews conducted with engineers and machine tool managers.

7 Company slogan: "Not simply selling machines but solutions!" (INT00890A)

8 Future technical analysis will determine to what extent this turns out to be a real capability or merely a marketing gag.

the German case: only 46% of the Japanese sample perceived the combination of machine modules as a means of customer influence. This figure gives a slight indication that buying off the shelf seems to be more common in the Japanese standard machine market, which is surprising while contradictory to interviews with representatives of Japanese machine tool companies. Further qualitative investigations will possibly clarify this presupposition.

These initial indications of relatively extensive influence on the part of customers, not only as a specific feature of German industrial culture but also to a substantially high degree in Japan, though possibly in contrast to other industrial cultures, will be dealt with in more depth below. This will include presenting statements regarding the origin of the decisive development impulses for CNC machine tools. In response to the question of the origin of the most important design impulses, sixty-five percent of the German sample indicated that they originated **internally**, fifteen percent considered them to be mainly externally induced and about eight percent saw both possibilities as having the same significance. The question concerning the importance of certain institutions as crucial external influencing agencies provided the following results (see Figure 4): the majority of the responses perceived the **customer demands** as the chief impulse for development of CNC machine tools (73%), **applied research** was regarded as the crucial factor by approx. 50% and roughly 33% considered the **knowledge of materials and processes** to be crucial. Science and basic research are seen as the major source for machine developments by only a small number of respondents.

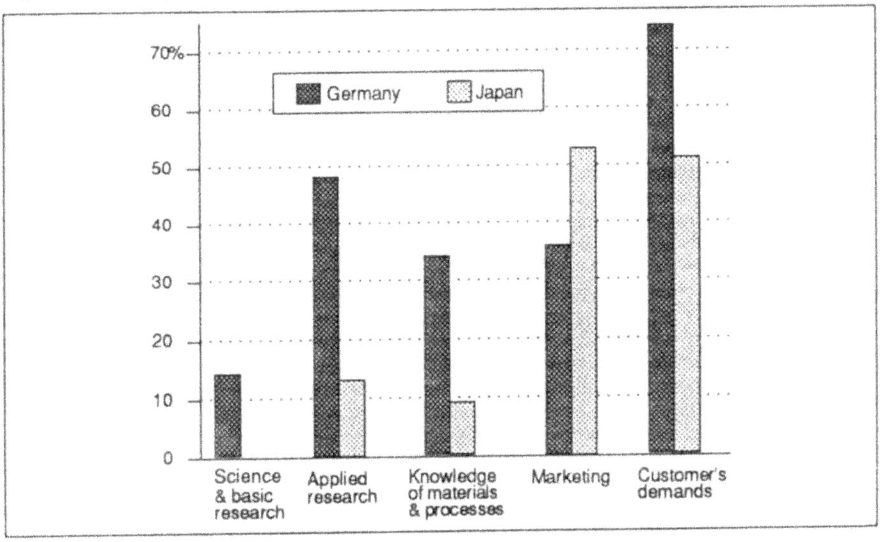

Figure 4: Origins of major Development Impulses
(Source: German and Japanese Machine Tool Survey 1990)

The quite considerable deviation of the Japanese sample with respect to this point is somewhat surprising in this context. It is true that customer demands are regarded as the source of decisive development impulses by approx. 51% of the responding companies, and this must be interpreted as a convincing indication of a high degree of customer orientation on the part of machine tool producers. However, a direct comparison regarding this item seems to confirm the widespread view that German manufacturers take into consideration customer demands as decisive development impulses to an exceedingly great extent. The relatively high proportion of responses in the Japanese sample indicating marketing as a major development impulse (53%) is also of significance. Two complementary explanations are suggested: first of all, the marketing department (insofar as such a department exists at all, which one can usually only assume for larger companies) is, so to speak, the interface between customers and the company; for the development engineers primarily addressed by the questionnaire customer demands are, as a rule, conveyed via the sales and marketing departments so that "indirect customer demands" can be assumed with regard to this point. Secondly, however, the size of the company certainly play a role, too: in the Japanese sample there is a relatively large number of big companies operating at overseas markets, for which marketing in general plays a more important role (as compared to smaller companies).

An analysis that differentiated according to company size in the German sample provided the following picture: only 9% of the companies having fewer than 100 employees indicated marketing as an important source of machine development while the figure for companies with more than 500 employees was over 50%. We can presume for both samples that customer demands (conveyed via the marketing department or directly) are of great importance for the development projects of manufacturers.

The relatively low Japanese figures for **applied research** and **material and process knowledge** require some explanation which I cannot provide yet because of the lack of comprehensive information. In the German sample very high figures were obtained, as expected, for **applied science** (approx. 50%) and somewhat lower ones for **knowledge of materials and processes**, though at 30% still far above the corresponding Japanese figure. The in-house, generally more application-oriented research activities received the second highest frequency, after customer requirements, as the most important source for design impulses. This item did **not** vary in its significance between large and small companies; thus we can assume **that customer demands and applied research represent the two major sources for design impulses**. Enquiries regarding the complex of "knowledge of materials and processes" provided indications that the production-related experience of manufacturing companies derived from their production and installation work are addressed, i.e. the machine tool manufacturers appear in their **second function** as (pilot) users of their own technology and make use of this function as an innovation resource (INT01190B). The latter does not seem to play such an important role in the Japanese sample (and possibly among Japanese machine tool manufacturers in general). The question of whether or not this is a specific feature of the industrial culture will not be clarified until after evaluation of interviews still to be conducted.

On actors

To provide a deeper understanding of company innovation processes and innovation make-up, I would like to briefly describe the findings on **actors and institutes** that are of importance for development activities (especially for CNC control). Figure 5 represents the results of the German and Japanese sample in summary form.

The answers to the question were allowed to vary between "no importance" to "great importance" on a four-grade scale. The figures do not represent the valid percentages to give an impression of the differing share of rejection in both samples.

With respect to the customers or customer demands, the two diagrams provide, first of all, a confirmation of the above mentioned results. Both the German and the Japanese samples show a similarly high degree of significance of customers as an influential actor in design and development processes (more than 50% in both samples ascribed great importance to customers). This finding becomes even more evident in the German sample if one includes the figures for informal producer-user relations: this item is assessed to be of "great importance" by approx. 30% of the respondents. Only 9% of the Japanese sample ascribe "great importance" in this connection, but some 30% give the producer-user relations "medium importance".

If we concentrate somewhat on the differences between the two samples, we see that the competitors seem to play a highly significant role in the Japanese case: 40% ascribe "great importance" and about 20% "medium importance" to competitors as important actors/institutions for development activities. The German sample produced substantially lower figures in this connection.

The high degree of Japanese orientation to competitors can be explained in part by the historical development of Japanese machine tool production. A country like Japan, which has undergone a rapid "catching-up" development since the 2nd World War[9] and has made a name for itself, especially as a "learning newcomer" in the machine tool production industry, closing the technological gap between the leading foreign competitors very rapidly and efficiently, naturally orients itself to competitors to a greater extent since this very competitor orientation provided the prerequisite for success.

In the German sample, on the other hand, competitors as a factor do not play such a significant role; however, an assessment of "great importance" for this item was indicated by approx. 30%, thus having roughly the same magnitude as informal producer-user relations, for example. The competitors, therefore, also play an important role in the German sample, but not to the same extent as in the Japanese sample.

9 In 1955 58% of Japan's domestic demand (in value) was still covered by imports from the U.S. Up to 1966 this figure had dropped to under 11% and since the 1980s it has decreased to a *quantité négligeable*.

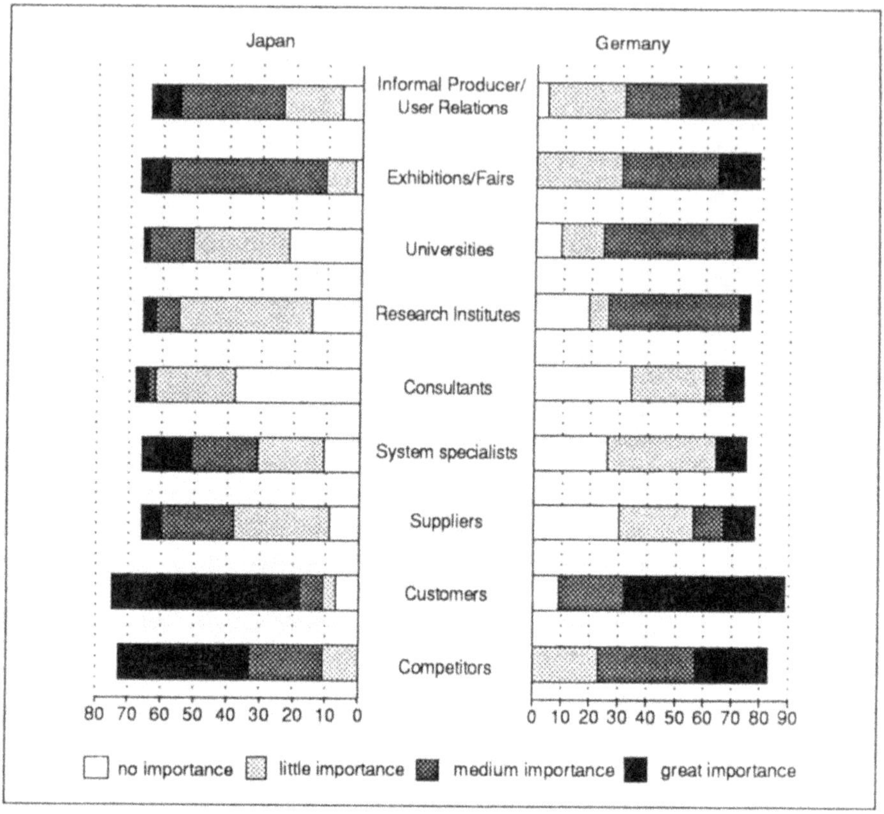

Figure 5: Importance of Actors and Institutions in Development Processes (Source: German and Japanese Machine Tool Survey 1990)

There is also a quite considerable gap between the German and Japanese sample with regard to the importance for development activities attached to *universities* and *research institutions*: less than 10% of the respondents in the German sample attach "great importance" to the latter, while "great" and "medium importance" together make up more than 50%. This reflects the large number of cooperative projects between manufacturers and institutes/universities, which are carried out, for the most part, on a commercial basis and thus are not dependent on state subsidies. The institutes include, for example, Fraunhofer Institutes as well as university institutes that have, in some cases, become specialized exclusively in machine tools and production-related problems. The cooperative relations are frequently of a long-term nature and have a long tradition (Pohl, Scheifele 1990). In the rather subjective judgment of a machine tool expert, particularly such cooperative efforts between manufacturers and institutes/universities represent the key to the rapid and successful linking of machine tools to electronics (i.e. CNC), which was of crucial importance in restoring the competitiveness of the German machine tool production industry at the beginning of 1980s and has since then contributed to constant sophistication (INT00990A).

In comparison, less than 15% of the replies in the Japanese sample ascribe great or medium importance (taken together) to universities and research institutes. This finding evidently stands in close relationship with the fact that the Japanese machine tool industry spent only 3.1% of its (accumulated) turnover on basic research (in 1990). In Germany, for example, the basic research expenditures of the machine tool industry figure slightly above 4% with large companies laying far above (e.g. Trumpf with 6%, Maho with 7.5% and Deckel with 6% in 1989) and smaller companies ranging markedly below the average[10] (*VDI Nachrichten* 34 and 43, 1991). It does not appear adequate to interpret this generally as evidence that basic research is a weak point of the Japanese machine tool industry. This problem complex requires further examination and interpretation (which I am not able to do due to lack of detailed information on the Japanese machine tool production industry).

The Innovation Make-up

If one pieces together the detailed information to form a meaningful picture of innovation make-up and processes, the following scenario for the German machine tool production industry presents itself:

In the overwhelming majority of cases, the manufacturers of machine tools regard customer demands as the most important source of development impulses. Depending on the market segment, customers exert influence via modularized machines (Maurice, Sorge 1990) in the case of more or less standardized machines and via performance specifications in the case of job-based series machines. In addition, manufacturers consider the informal producer-user relations to be another very important transmission mechanism for influencing or transmitting technical modification wishes and thus also for exerting technical influence on machine design.

If we include further sources of development impulses (applied research, knowledge of materials and processes) and the actors that more or less clearly correspond to them, then we get a constellation sketched in Figure 6: The typically rather incremental innovation processes are supported by an innovation make-up that is basically composed of three in-house areas/departments: the development and design department, the sales and marketing department (consisting of after-sales service and marketing) and manufacturing (i.e. in-house production and assembly).

The most important impulses originate from the sales/marketing department and customer services. Since one can interpret both departments as the hinge to the customers, via which the latter convey their wishes and experience to the machine tool manufacturer, the sales and marketing department is an important, if not the most important, source of impulses for technical development projects. Customer wishes, customer behavior as well as the investment strategies of the industrial branches supplied by the machine tool industry thus influence the

10 The European average is estimated to account for 4% (WS Atkins, 1990).

innovation processes, directions and speed of technical developments in the machine tool sector.

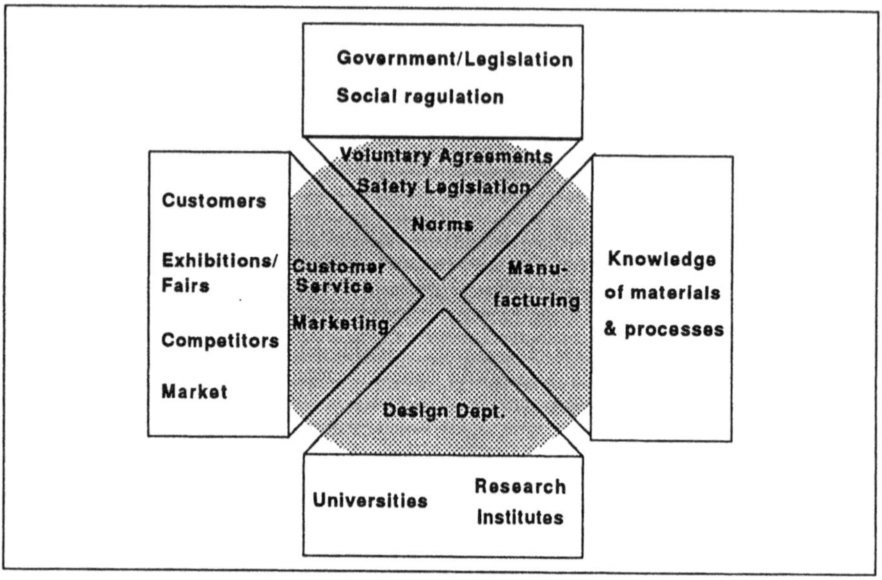

Figure 6: Innovation Make-Up in the Machine Tool Industry

The second important source of information are, of course, the development departments, where in the end all design impulses come together and are implemented in design processes. Furthermore, they also act as intermediaries for new, frequently more basics-oriented but, nevertheless, practice-relevant knowledge via cooperation with institutes and universities and thus, with new know-how, provide access to the development of new machines.

The third source of technological developments in the machine tool industry is in-house manufacture (production and assembly) - in this way the machine tool manufacturers have their own prototype tester. Many suggestions for products/innovation originate in experience gained on the shop floor, especially production-related development details (INT01190A).

The last complex of sources for design are norming activities, government's regulating activities (especially safety regulations, etc) and governmental r & d policy. The latter evidently can not determinate but offer incentives for certain technological and industrial-structural developments. Thus, governmental technology policy can support/accelerate future-oriented developments in certain industrial sectors or technological areas, e.g. the German programme on manufacturing technologies (shop floor oriented programming WOP; etc).

Taken altogether we get an internal-external coalition which means that company-internal actors or departments have external counterparts (actors, institutes, etc.). Both sides are formally or informally interlinked. Especially the informal relationships are highly dependent on favourable communication structures: At the intra-company level between the manufacturing area and the design department (with little "communication barriers") and with externals such as customers, suppliers, etc.

Interim résumé

The special features of German industrial culture or the specific production culture of the machine tool industry seem, therefore, to substantiate, as expected, the dominance of customer-induced development impulses in connection with the correspondingly emphasized significance of the departments maintaining close contact to customers. The great importance of the development department was, of course, also to be expected; however, we did not expect that special weight would be given to universities and production-related research institutes to this extent. A direct comparison with the Japanese sample signifies a clear difference on that point, comparisons with other countries will clarify whether this is a special feature of German industrial culture.

The utilization of the company's own production department as a prototype testing field for new machines and as a source of production-related experience which is inputted into machine developments seems to us to be a special feature of the German industrial culture, by means of which the production know-how of skilled workers is translated into detailed modifications of existing machines (INT00990 and INT011A).

In the situation described standardization activities and legal regulations play a rather subordinate and probably culturally indifferent role, in contrast to state-initiated technology promotion activities, such as WOP, which are strongly bound to the industrial-cultural context and which owe their existence and their success to this thematic coherence.

Let us now take a very brief look at the consequences of this innovation make-up for the competitiveness of the German machine tool industry. The extreme customer orientation is certainly an efficient means of achieving incremental detail improvement, which is indicated by the records of concrete deficiencies in existing machines. This is undoubtedly the basis for a high degree of adaptability and flexibility in the German machine tool production industry. Future developments concerning machine tools as well as CNC controllers thus grow, so to speak, out of the current problems in practice and less out of research activities of the development departments. This is, however, a limited and, in the end, defensive innovation perspective, which corresponds very precisely to the pattern that is comprehended in economic theories of innovation as **demand-pull dynamics**[11]. These approaches see technological developments as mainly induced by the demand for certain technologies. The manufacturers' task is to perceive the demand and offer products/technologies which meet the demand. This constellation describes exactly what happens in the customer-manufacturer relationships which we characterized as a German pecularity above.

---

11 At this point we do not want to go into the still undecided debates on the pros and cons of demand-pull and technology push, respectively.

Although technology push structures do exist, if one looks, for example, at the cooperative relations between development departments and universities/research institutes, these dynamics are considerably weaker as compared to the demand-pull forces.

The possible future threats for the competitiveness of the German machine tool industry could be seen in the fact that dominant demand-pull structures, formed via the industrial culture, along with the customer demands as central agent show, overall, an exceedingly weak degree of dynamics, that they can "only" induce incremental developments. Though this does contribute to the highest degree of sophistication for traditional production methods (such as high-precision, high-speed turning), **it is probably not sufficient for pioneering development of new methods** (such as laser-aided turning, machining or processing of new compound materials, etc.), **new materials and tools and could, therefore, prove to be too weak for future requirements.** Thus it is to be feared that customer-oriented innovation strategies are not powerful enough to exert a fundamental influence on innovation processes, i.e. to determine the innovation speed and directions. The conclusions of the *Strategic Study on EC Machine Tool Sector* also point in a similar direction. The latter expresses doubts when it states that the EC machine tool companies lag behind their major competitors regarding r & d in the application of new materials, mechatronics, laser technology, etc. due to the small average company size and thus correspondingly low level of financial and human resources in r & d (WS Atkins, 1990, p. 21).

At this point it would be very interesting to conduct a direct comparison with the Japanese and other national machine tool industries and look at how their innovation processes and capabilities are shaped. If we leave the actors and sources of innovation and concentrate on the substantive orientation of the innovation processes, we have to consider the design orientation that underlies certain developments of machine tools and CNC controllers, thus returning to one of the central items in industrial culture: *skilled machine work.*

4 Dominating design orientations

In our hypothesis regarding the design orientation for CNC machine tools we assumed that skilled work is an important reference point for developments of mechanical engineering and machine tool development in particular. The industrial culture concept presupposes that in the German case the vocational training system and its outcome "skilled workers" (Facharbeiter) accounts for rather more work oriented CNC control systems, which are rather more geared towards skilled machine workers. We assume **that machine design and the design of CNC controls is not strictly determined by technology but to some extent geared towards the potential users, their capabilities, skills and knowledges.**

Presumably there exists an industrial cultural interference between the problem solving perspectives of engineers and the prevailing production concepts, work organization and utilizing conceptions for machine work. To find out supposed correlations between work oriented machine design and the prevailing qualification structure of machine work a brief statement on the general development and production strategy of the German machine tool industry might be helpful. With respect to the question of single-process orientation the German sample is in accordance with our expectations: Approximately 80% of the weighted sample produces machine tools mainly for one process (we defined a share of 70% or above as the crucial size). More than half of the sample lies markedly above the figure and those gathering around the 70% borderline combine either one process with the production of FMS or two similar processes (such as surface grinding machines and bed-type milling machines, etc). There was only one pronounced warehouse producer manufacturing lathes, grinding and milling machines as well as FMS in almost equal shares. As far as quanitative data allow a conclusion here it would be that the companies in the sample are concentrating on one process and thus are accumulating and developing knowledge in their area. Questions raised in the face-to-face interviews on that point caused corresponding answers in the way that the companies' philosophies say that a machine "has to be growing out of the process logics" and not following the CNC hard- and software logics (INT01190A). So the process dedication seems to be a first real strong orientation of the German Machine Tool Builders.[12] As a consequence of the pronounced single-process orientation it can be expected that the CNC control systems combined with the machine tools are to a large extent process-specific.

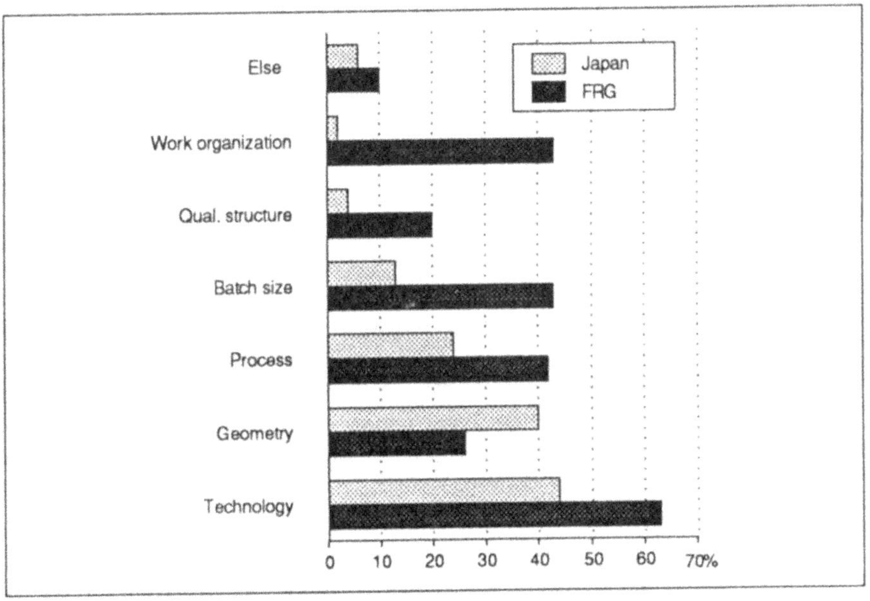

Figure 7: Reference Factors for Design and Selection of CNC-Technology
(Source: German and Japanese Machine Tool Survey 1990)

12 Unfortunately, a direct comparison with the Japanese sample is impossible, because we only had the processed data (frequencies, means, etc) available and thus could not conduct a casewise evaluation.

In the German sample process and technology orientations are the major reference factors for design and selection of CNC (see Figure 7). This finding confirms the above stated machine-dedication to one process and furthermore gives hints on non-technical orientations. Thus we get two groups of reference factors, one which is relatively independent of the companies' field of operation (comprising of technology, geometry and process) and another group of factors which are highly sensible for the operational field. Two dimensions of the latter have an outstanding significance:

- the work organization in the customer companies and
- the qualification structure in the customer companies.

The peculiarity in the German case seems to be the general awareness of work organization and qualification structure as a referential system for CNC design. On the contrary, the Japanese sample attaches almost no importance to these factors. By conceding significance to work organization and qualification structure we do not know much about the substance of these factors: Does the awareness result in adaptation towards the customer companies' prevailing qualification and organization structure; and what is perceived as the dominating pattern of qualification and organization in the machine tool utilizing companies? Is it a more Tayloristic, technocentric or anthropocentric general orientation which is supposed to be the prevailing structure of machine tool applying industries?

The task and operation relatedness of CNC machine tools also comprises of more implicite qualification and organization factors, which, for the German situation, clearly means an **orientation towards skilled work** as the dominating form of utilizing machine work.

User orientation

In our survey we asked developers in machine tool companies which group or category of CNC-machine users they anticipated to be the user/programmer of their CNC machine tools (see Figure 8). As was to be expected, fifty-four percent of the replies in the weighted sample marked *skilled operators* (Facharbeiter)[13] and an additional eleven percent claimed *unskilled or semi-skilled operators* as the expected user group. We expected this because of the relatively large amount of shopfloor-programming systems occurring in the sample. What we did not expect is the fact that fifteen percent replied that their CNC-machine tools can properly been programmed both in the programming bureaus as well as on the shopfloor (about fifteen percent claimed qualified programmers in programming bureaus as the appropriate group). In accordance with our assumption that there is a correlation between company size and the expectation of skilled workers as the potential machine programming group we calculated a $\lambda_{machine\ prog}=0.6$. This indicates that the smaller the company the higher the probability that skilled work is the expected form of machine operation (programming particularly).[14]

13 The unweighted sample figured about 40%.

14 The $\lambda$was calculated at a significance level of 95%.

In a complementary question we asked whether the CNC in question supports the transfer of programming and disposition to the shopfloor level or to the programming and planning bureaus. Nineteen percent replied *bureau* and twenty-three percent mentioned *shopfloor*. The interesting fact is that fourty-six percent saw both possibilities. Thus the skilled worker is not the exclusive but a very important **user model** for the designing engineers. Face-to-face interviews with engineers confirmed the finding. One engineer of a medium-sized machine tool company put it like this: "A machine that cannot be programmed by a skilled worker (Facharbeiter) on the shopfloor **and** in the programming bureaus is **useless** (INT01190B)."

Together with the above sketched item both factors indicate a strong orientation towards skilled work **and** at the same time a kind of versatility and organizational flexibility that is covered by the CNC machine tools. Openness towards the needs of the SMEs (e.g. CNCs which consider the low levels of bureaucracy and the the little capacities in work preparation, etc) as well as to the different requirements of mass producers are both substantial cornerstones.

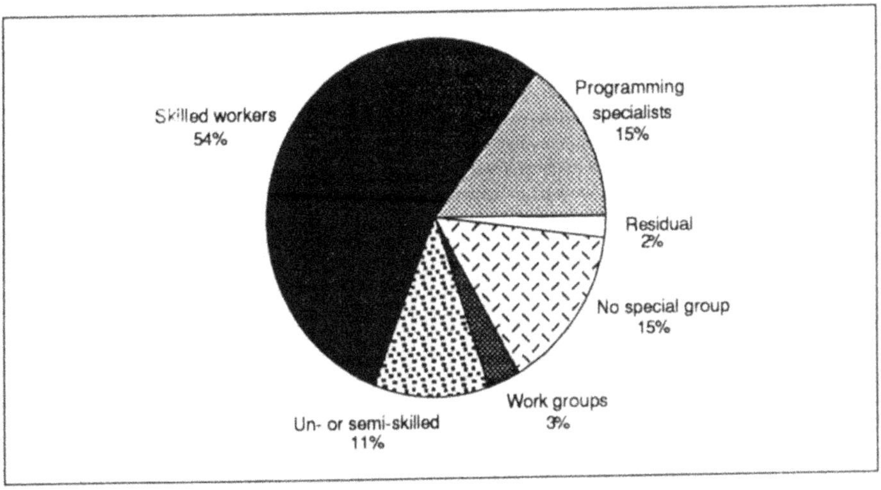

Figure 8: Images of Potential Users/Programmers
(Source: German Machine Tool Survey 1990)

Recent discussions seem to modify the argument at one point (which was not in the centre of our survey): Beyond the appropriateness of CNC controls, the machine tools are often over-sophisticated and technologically too elaborated for the needs of SMEs (Politsch 1991). Japanese machine tools often are of greater simplicity and thus are better adapted to the technical and financial requirements of SMEs.

The Japanese sample differs from the findings in the German case (see Figure 9): The dominating image of potential users/programmers are programming specialists (26.7%). Japanese programming specialists are contrary to German programming specialists ferquently located on shop floor level (in Germany specialists are expected to work in the work preparation and thus are *far away* from shop floors). Skilled workers are expected to be the potential users/programmers in 13% of the cases.[15] and similarly to the German sample unskilled or semi-skilled are of minor significance (they figure about 2%).

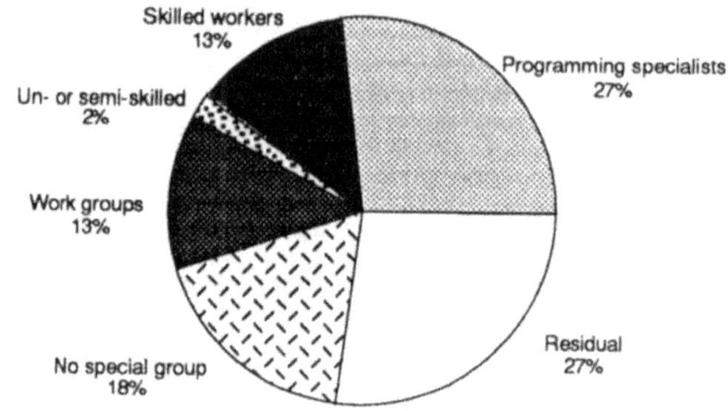

Figure 9: Images of Potential Users/Programmers
(Source: Japanese Machine Tool Survey 1990)

Interestingly, 13% of the Japanese sample ascribed significance to work groups as potential users/programmers. Evidently this orientation refers to the widespread institution of **work teams** in Japanese industry (cf. Womack, et al. 1990). In contrast, the German sample gave only peripheral significance to the item (3%). One obvious reason is the relatively little occurence of work groups in the German metal processing industries.

To get a clearer picture of the dominating orientations in the design of CNC machine tools, it makes sense to extend the analysis on the goals that machine tool manufacturers (here: design engineers) associate with the CNCs developed. Beforehand we should realize that approximately 60% of the machines described in the sample were assessed as **suitable for shop floor programming.**[16] The orientation towards shop floor programming is no contradiction to the common paths of factory automation. Many German machine tool builders nowadays design CNC machines that meet both requirements, the shop floor programming and the ability of integrating the machines in CIM process chains. Interviewees clearly labelled it as a **parallel strategy**.

15 Evidently, in a formal sense skilled work is not comparable with the meaning it has in the German context, but measured in terms of qualification level the substance is probably similar in both cases.

16 The assessment did not rely on a self-assessment of the machine tool manufacturers but was based upon assessment criteria determined by the project team. The criteria used to assess the shop floor programmability were assigned to the following areas: shape of user surface; dialogue control (interactive, etc); program testing support (plausibility check, etc); adaptability of menus and functions; etc.
In a control questions we asked the manufacturers for an assessment whether or not their CNC is fit for shopfloor programming. The answers confirmed our assessment.

On Design Goals

The great advantage of a technology that is to a large extent geared towards the shopfloor operations must be seen in using the flexibility of organization which itself can be translated into a flexibility of production. The following Figure 10 gives an overview of the respondents' assessment of the degree to which their machines fulfill certain goals (the questionnaire provided 4 ranks from "largest extent" to "not at all"):

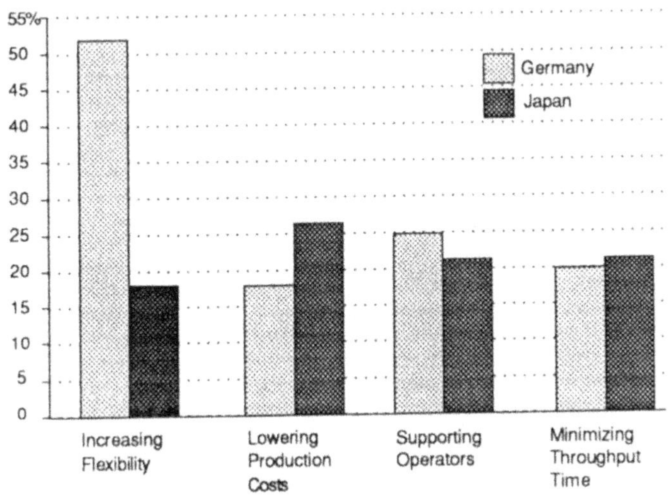

Figure 10: Goals fulfilled by the CNC-Machine Tools
(Source: German Machine Tool Survey 1990)

The figure maybe says not so much about the real characteristics of machines and controls but it says a lot of the rank of production goals that are followed. The unrestricted leading goal is the *rising of flexibility* which is assessed as extremely important by fifty-two percent (if we add the category "to a large extent" we get a figure of about seventy-four percent). *Supporting the operators* is perceived as a goal which 25% of the sample believe to be fulfilled by their respective CNC machine tools to a largest posible extent. *Minimizing throughput time* follows with 20%. Surprisingly "lowering production costs" is not recognized as a major fulfilled goal (less than 29% believed it to be fulfilled to the largest extent possible) – a contradiction to the expectations of the customer companies.

If we now add the data of a related issue on the relative importance of certain factors for an ideal/optimal/successful machine design we get the following picture: In the view of the respondents a successful machine seems to be the one which meets best the technical requirements (fourty-eight percent indicated extremely important, the highest rank out of six). The most interesting answers were given to the item *fulfill the requirements of users*. Here about fourty-five percent of the respondents indicated "extremely important", the highest rank and another fourty-five percent indicated "important". The respondents seem to be strongly oriented towards the users which in their view should be a high-priority goal in the design of an ideal, successful machine (or system). In addition to the mentioned item another stimulus is important. We asked for an assessment of the postulation that an ideal machine design should utilize **and** enhance the skills and qualifications of the machine operators. Twenty-five percent assessed it as extremely important, thirty-seven as important and twenty-nine indicated "not important".

In the German sample *supporting operators* and *enhancing skills and qualifications of operators* both correlate to the relative share of shop floor/manufacturing personnel in the machine tool company: The higher the percental share of manufacturing staff (which is outstandingly high in small companies), the higher is the consent to support operators and enhancing their skills and qualifications ($\gamma_{support}$=0.50; $\gamma_{enhance}$=0.53) or, in other words, the stronger is the projection of the machine tool builders' organization and qualification structure towards the structure of the customer companies.

As a **result** we get a specific paradigm of factory automation which is constituted by the industrial-culturally influenced design strategies and problem-solving perspectives of engineers. The high rating of the fulfillment of *user requirements* as well as the relatively high scoring of the demand for *utilization and enhancement of operator`s skills* indicate the outstanding significance of the prevailing work and qualification structure in the German metalworking sectors as a cornerstone for design and thus can very carefully be interpreted as a more work-oriented, anthropocentric path in machine and systems design.

With regard to the goal fulfillment the Japanese sample shows similarities with the German one. The consent to flexibility as an important goal to fulfill is not as strong as in the German case (even if we count together the categories "largest extent possible" with "to. a large extent" we get 47% which is less than the German figur for the highest rank) but the support of operators is assessed with a similar figure as in the German case. If we take the related item on the important factors for an ideal/optimal machine design we get the following picture: About 50% estimated the fulfillment of *practical and technical requirements* as the most important factor (similarly the German figure was about 48%). The similarity continues with the item *fulfilling the user requirements* which, in the Japanese case, was seen as the most important factor by 44% of the respondents. A significant difference between the two samples occured with the postulation that machines ideally should utilize and enhance operators' skills: Only 16% of the Japanese respondents assessed this item as "extremely important", 27% as an "important" element of an optimal machine design. Thus, the skill enhancement is not valued as highly as in the German sample. Instead we get a clear preference for the postulation that an ideal machine design should meet the market demands (33% estimated this as extremely important). The Japanese sample emphasizes the technical and market-oriented goals (similar to the German sample) but there is no such pronounced orientation of machine design towards supporting operators and enhancing skills and qualifications of machine operators as is in the German case.

5 Conclusion

Comparing the innovation capabilities, processes and typical innovation set-ups of the machine tool industries in Germany and Japan we find beyond some similarities striking differences which may take crucial significance for the future competitiveness of the industrial sectors in question.

In both countries the customers of the machine tool branch use the different means of interference in a differing intensity. Machine module combination in the segment of standardized CNC machine tools and requirement lists as well as the far reaching means of joint conceptual design in the more customized market segments are common in both industrial cultures. Comparing both country surveys gives strong hints that in the German industrial culture the customers' influences play a more important role as in Japan: the customer demands seem to be the key element for the technological development of machine tool manufacturers and they are furthermore corresponding to the intra-firm innovation efforts on applied science and process knowledges.

The customer-driven more application-oriented innovation processes follow **demand-pull dynamics** - this is also true for the Japanese case, but on a supposedly smaller scale as in Germany - which are complemented by weaker but still remarkable technology-push dynamics (originating in the cooperation between r & d departments and externals such as institutes and universities). In the considerable absence of basic research in the Japanese machine tool industry lies the probably most significant difference to the German case <u>and</u> even more important this deficiency might turn out to be the major threat to the future international competitiveness of Japanese machine tool industry. This danger evidently also threatens the German machine tool industry. Thus, our findings indicate that the demand-driven innovation processes result in machine designs which are highly **appropriate to the industrial cultural context** in which they are embedded and where they originate from. The adequacy is twofold: The quality of the developed CNC technology meets the requirements of small and medium enterprises (SMEs, which are dominant in the German metal-working industry), their prevailing skill and qualification structure and on the other hand **the demand-driven innovation processes are *the innovation processes highly appropriate to small and medium machine tool manufacturers*, appropriate to their (financial and human) resources**. The obvious thing that suggests itself is the formal (emphasis on formal!) analogy to the early NC developments in the U.S. which were directed by a strong demand-pull from the aerospace and later also by the automotive industry with the result of a ristricted development perspective by the machine tool manufacturers. A similar trap contains the contemporary innovation make-up in the German industrial culture. Thus, the dialectics of the story is that as far as trend-setting technological challenges are to be overcome (such as new techniques, new materials or ultra-precision machining, etc) customer demand-driven forces are not sufficient and stronger technology pushs are required.

Since the smaller machine tool companies have almost no means to increase their financial and human resources their chance lies in strategic collaborations with competitors, institutes and suppliers. The major German consulting companies suggest to build up networks of up-to-date large-scale enterprises or to merge (i. e. the "Japanization" of the German machine tool branch structure) or holding companies to increase the competitiveness of the german machine tool industry.

An improvement of the national innovation system, in particular the innovative infrastructure is required to enable the smooth transfer and mediation between basic research, applied research, process and material knowledges and users' requirements.

References

American Machinist, various editions.

Häusler, J. (1990). *Zur Gegenwart der Fabrik der Zukunft: Forschungsaktivitäten im bundesdeutschen Maschinenbau* (Translation: Contemporary Factory of the Future: Research Activities in the German Machine Tool Building). MPIFG Discussion Paper, Köln.

Maurice, M. and A. Sorge (1990). *Industrielle Entwicklung und Innovationsfähigkeit der Werkzeugmaschinenhersteller in Frankreich und der Bundesrepublik Deutschland* (Translation: Industrial development and Innovation Capability of the Machine Tool Industries in France and the FRG). WZB Discussion Papers, Berlin.

Pohl, W. and M. Scheifele (1990). Eine Forschungsgesellschaft im Umbruch zum Konzern. In: *Wechselwirkung* 45/46, 1990, p. 61-67. Berlin.

Politsch, H. W. (1991). Die Zukunft liegt im Kleinbetrieb. In: *Technische Rundschau* 6/91, p. 16-23.

VDI Nachrichten, various editions.

WS Atkins Management Consultants (1990). *Strategic Study on EC Machine Tool Sector.* Executive Summary. Brussels.

Womack, J. P. et al. (1990). *The Machine That Changed the World.* New York.

Shop Floor Oriented Programming - Experience from a Joint Development Project with the German Machine Tool Industry

by Peter Brödner

1 NC Technology: Intelligent Machine versus Computer-aided Craftsman

Past development of manufacturing can be characterized by three main stages. At the first stage, the holistic work of traditional craftsmen has been horizontally divided into subtasks and specialized activities according to the principles of Smith and Babbage. At the second stage, Taylor's principles of "scientific management" introduced vertical division of labour separating conception from implementation. At the third stage, this vertical division of labour has been further extended by introducing numerically controlled machine tools, as programming has been made a special task of work planning.

This development process has been driven forward by the socio-economic imperatives and prospects of more effective control over production processes and of improved productivity rather than technological necessities. With this respect it was Taylor's declared intention to stop the suspected "idling" of the workers, to take on "the burden of gathering together all of the traditional knowledge which in the past had been possessed by the workmen and then of classifying, tabulating and reducing this knowledge to rules, laws and formulae", and to put it in place of the old "rule-of-thumb methods" and the "workers' discretion" (Taylor 1911, 36).

The application of these principles soon revealed some severe difficulties and obstacles, however, in particular in customer-oriented manufacturing. Here products and production processes are, in many cases, so complex and diverse, flexibility and quality requirements are so high, processing knowledge and situational comprehension are so limited that production processes cannot be analyzed and modelled precisely and completely enough to derive sufficient plans and prescriptions for production. These principles, thus, formed a programme rather than a realistic desription of manufacturing practice (Brödner 1990).

The introduction of NC technology and, more generally speaking, of "programmable automation" then promised to form a new base for applying the Taylor model to flexible manufacturing and to achieve a quantum leap in approaching the overall objective of getting independent of the workers' skills and self-will.

Although this again turned out to be rather illusionary, since in most cases effective and efficient manufacturing could not be done without skilled workers, the Tayloristic view preserved its dominance until recently: "German craftsmanship and the well known quality work of German skilled workers will no longer be needed in the future. They will be replaced by flexible systems and intelligent control." Statements like this one by David Grossman of IBM (Landesreport 1987) mark the common belief that knowledge based systems and "artificially intelligent" machines will finally succeed where traditional techniques failed so far.

There are mainly two reasons why this view is mistaken. The first one is objective in nature and relates to the fact that there exist severe difficulties and fundamental limits of converting the practical experience and tacit knowledge (knowing how), which is at the core of skilled work, into explicit theoretical knowlegde (knowing that), which is needed for automatic production (Brödner 1989 and 1990, Böhle and Milkau 1988). The second one, more factual, points to empirical evidence that flexible production and the efficient use of NC technology highly depend, although to different degrees in various countries, on skilled work. Despite the global effectiveness of the economic imperatives of capital exploitation, considerable and persistant differences in work organisation, in the concrete shape of technology and in the way how it is used remain observable in different industrial cultures.

The development and use of technology are a result of social relations, interests and needs. More, better, faster and more independent, these are the socio-economic imperatives of manufacturing under competition that determine the general movement and direction of technological development. The concrete shape of technology, the division of functions and the forms of interaction between humans and machines, depend, however, on the understanding and conceptions of the role of humans in production. Moreover, the technology produced, in turn, sets the conditions of acting for the operators to use it productively. Depending on the specific design decisions taken, the new working situation, work organisation and skill requirements will be more or less productive, demanding and attractive, and it will affect attitudes towards work. These are the basic reasons why the course of development and use of technology is determined by cultural factors such as values, traditions and models as well. This can be verified for production models and processes in general, and it is in particular true in the case of NC technology. It can serve as a demonstration how conventions, traditions, values and visions superimpose economic driving forces and influence the development and use of technology (Rauner and Ruth 1990).

A short look on the development and use of NC technology in the USA, Japan and Germany reveals the relevance of industrial cultural determinants modifying the economic imperatives of cutting costs and improving quality and productivity under global competition. This comparative perspective shows considerable differences in conceptions of the function of machines and the role of humans in manufacturing.

The USA have for a long time been the leader in manufacturing productivity (in particular with respect to mass produced standard goods), they have been the leading machine tool producer and the pioneer of NC technology. Despite these advantages they experienced a considerable set-back in production competence and industrial competitiveness more recently. Management's persistent attempts towards total control of manufacturing and replacing direct labour by programmable machinery control of manufacturing and replacing direct labour by programmable machinery brought the remaining work force into a situation of inattractive and less demanding work where little opportunities to develop new skills were left and existing skills were wasting away. This, in turn, enforced the development of capital intensive high-tech programmable automation equipment for special needs and branches (e.g. aerospace, military complex, cars), while in many other manufacturing industries, due to the losses of production competence and poor productivity gains, processes of self-destruction got going. The decline of the machine tool industry, which, as a key industry for manufacturing, also impairs other branches, clearly indicates this de-industrialization.

In Japan a different view and understanding of technology, seeing machines as means to make skilled work more productive, is dominant. NC technology has been developed, from the beginning, in terms of reliable, low cost, easy-to-use machines that served the needs of big and small enterprises as well. Available figures show that from the late seventies onward almost two third of inner Japanese NC machine sales have gone to smaller enterprises with less than 300 employees. The use of NC technology was combined with skilled work, technological skills and technology could develop together. This fastly growing demand induced, in turn, the production of standard NC machines. With exports not higher than 30 %, the rise of the Japanese machine tool industry can only be explained by these internal Japanese market relations and development conditions.

In Germany, NC technology was first of all adopted within the Taylor model of production, also with the declared intention to replace skilled work. This led to high frictions with the skilled workers, however, who were, after all, being regarded as indispensible for operating the expenisve equipment, but found themselves in very unfavourable working conditions with little influence on the processes. With the advent of computerized NC technology (CNC), a new line of development started, in particular with the needs of small or medium sized enterprises in mind, which allowed the workers to program or to adjust programs on their own and, thus, to preserve their skills.

This is, very roughly sketched, the context, in which the development of shop floor oriented programming methods is embedded. It is outlined in the next sections.

2 Looking back: Development of NC Technology in Germany

The beginning of main stream development of NC technology in the USA was based on measurement, control and memory equipment being created during World War II. The basic idea was to define a method for exact prescriptions of tool movements of a machine tool in numerical form and to automatically coordinate the machine cinematics using feedback from measured real positions. The target application was to produce surfaces with arbitrary curvatures for airplane parts. There was a vested interest of the US Air Force to reduce the enormous efforts for the creation and processing of the geometric data needed and to control the machining process independently of the workers' skills. The result was a problem-oriented programming language called APT (Automatically Programmed Tool). It focused on the definition of arbitrary geometric shapes of parts and the automatic processing of geometric data for the machine tool movements. The whole development was dominated by the interests of the US Air Force in an unified, general, but still flexibly applicable programming method. The Air Force set the specifications, organized the standardization, and subsidized the first application, thus actually creating the "market" for this technology (Noble 1984, Shaiken 1980).

At the same time, an alternative line of development with a different technical and organisational orientation also existed. This "record-playback" technique, being initiated and promoted by Gisholt and General Electric in 1946/47, provided the first part of a lot to be produced by a skilled worker while the signals from machine movements and states were recorded on magnetic tape. These signals were then used to control the machining process for the next parts of the same type. Although this playback technique was offered as a technical option in GE controls for a long time, it found only little resonance with management still looking for replacement of skilled work (Noble 1984).

The APT development was picked up and modified later on by German machine tool builders and research institutes. The modifications, layed down in APT derivables like EXAPT, AUTOPIT and others, focused mainly at two development objectives: simplifying and reducing the definition and processing of the geometric data for normal engineering parts and extending the software towards automatic determination of technological data like cutting speed, feed and depth. Since, at one hand, APT was designed to handle the most complicated tasks of 3D-continuous path manufacturing of surfaces with arbitrary curvature, it appeared to be rather inappropriate and too costly for normal point-to-point operations (boring) or 2D manufacturing tasks in turning and milling. On the other hand, the German development aimed at expanding the scope of the programming system by automatic processing of technological data. Therefore, the further development was necessarily operation-specific. With this intention to further automate NC programming through automatic determination of cutting parameters (use of a technology processor in addition to the geometry processor), the German development followed again the Taylor model of separating conception from execution (Stute 1969). NC programming became part of work scheduling (whereby even part of the scheduling work was objectified).

The skilled workers remaining on the shop floor now found themselves with poor and unsatisfying working conditions. Since they were confronted with NC programs made by work schedulers in the technical office, their occupational skills and intellectual abilities were hardly demanded any more or seldom called off in problematic situations. The typical NC operator work was characterized by abstract reliability requirements, unspecific and simple monitoring tasks and one-sided work loads connected with load or unload operations. They typically suffered from stress through too little demands. They had continuous conflicts with the NC programmers in the planning office, as the programs, being conceptualized far away from the real machining processes and lacking the background of practical experience, normally contained many errors. The workers were then asked to help correcting the programs which they were not allowed to produce, but which they could have produced with less errors and higher efficiency due to their practical experience (as empirical evidence corroborates again and again). This situation resulted in high interface losses with long set-up times at the expensive NC machines and big assignment problems with high fluctuation of workers. The existence and interests of a skilled work force, the dominant production model and the shape of technology were in obvious conflict, economies and politics of production contradicted each other.

As a consequence of this, real cost reductions disappointed the expectations and the diffusion of NC technology was more or less restricted to larger companies with Tayloristic work structures and well established planning offices. The demands of the many small and medium sized companies, although steadily gaining economic importance, were rather bypassed by this line of development. In the middle of the seventies, diffusion rates of NC-technology stagnated in Germany. It was not until the advent of CNC technology with storage programmable in place of hardwired programmed controllers that the diffusion experienced a new upswing. The new technical flexibility in defining control functions through software facilitated a number of input devices for the skilled workers on the shop floor to correct or even directly conceptualize and produce the NC programs. These opportunities also opened the doors of smaller companies for NC technology. In accordance with this change, NC machine sales increased by leaps and bounds: while in 1974 only 4.000 NC machines were in use, the stock went up to 25.000 in 1980 with growing rates per year (Rempp and Lay 1981).

As a consequence of this, the spectrum of ways how to use NC technology in production widened considerably. It stretched from strict office programming (according to the Taylor model) to pure shop floor programming through the skilled workers (with several intermediate forms). This is mainly decided according to the organisational model, personell development and assignment policy a company has rather than technical factors (Rempp and Lay 1981, Sorge et al. 1982, Kern and Schumann 1984). This situation gave rise to a dispute on organisational models being fought in kind of a religious struggle between the advocates of shop floor and office programming. Empirical evidence undoubtedly showed, however, that, under a diversity of conditions being typical for customer-directed manufacturing, NC programs are produced much more economicly efficient through the skilled workers on the shop floor than in the technical office. Programming costs are less and program efficiency is higher (e.g. with less run time by 30 % measured in case of turning) leave alone the considerably reduced set-up times for testing the programs. These advantages directly relate back to the fact that the practical experience and situational knowledge about the actual states of tools, fixtures and machines, being available on the shop floor only, are comprehensively used (Lay et al. 1983, Nyhuis and Nyhuis 1986).

A recent phenomenological study of skilled labour in connection with the use of NC technology confirmed again that, despite the partially automatic operation of NC machines, skills and tacit knowledge of production are still being needed on the shop floor in order to use them efficiently. Sensual perception and experience is an essential part of working with them. It is held that not all practical experience of production, and of the cutting process in particular, can be explicated or transformed into objectified scientific knowledge, because of its "tacit dimension". The sensual experience of skilled workers as a different form of perceiving reality can be characterized by the holistic sensual perception of the cutting process, the emotional involvement of the workers, the personal relation to the machine and by the fact that workers normally don't follow rules (Böhle and Milkau 1988). Unless the operators are also charged with programming, tool preparation, quality control and routine maintenance, their adequate holistic reality perception would be hindered. They would have great difficulties in acquiring and maintaining the practical experience.

At the same time it was more and more realized that traditional abstract programming methods and editing devices of modern control units do not meet the modes of thinking and working of skilled workers and are, therefore, hardly suitable for them. They know to read drawings and to transform their contents into cinematic motions of the machine tools. Their knowledge is, as shown above, essentially based on sensory experience and, therefore, vivid and analogic. Traditional programming methods, however, stem from the realm of engineering and data processing and are abstract, algorithmic and digital.

A new perspective in system design was needed to develop programming methods being more adequate for the shop floor.

3 The Development of Shop Floor Oriented Programming

In the late seventies a number of development efforts had taken place with this new design perspective in mind. They resulted in much more appropriate editing devices of control units, in a process-oriented interactive graphical programming procedure based on pre-programmed, parameterizable cutting cycles for the turning process, and in graphical dynamic simulation of programs (Diekmann and Klotz 1980, Spur and Potthast 1981, v. Zeppelin 1981). Encouraged by the very positive experiences, an industrial working group was formed by the initiative of the Federal Manufacturing Technologies programme management agency in order to specify the functional requirements for a shop floor oriented programming method suitable for skilled workers (Liese 1983).

Based on these results, experiences and requirements, a joint R & D project with a number of machine tool builders, control and software suppliers, research institutes and a pilot user was established within the framework of the Federal Manufacturing Technologies Programme. The objective was to develop new shop floor oriented programming methods suited for the different processes of turning, milling, grinding and sheet metal cutting. Despite the necessary process specific features they provide a unified user interface giving the skilled workers computer assistance as a tool at hand. It allows them to make use of their manufacturing knowledge, their planning capabilities and their situational knowledge about the states of machines and tools in order to efficiently produce correct and optimized NC programs. By doing so, they can also reproduce their competence.

Accordingly, the functions and tasks are allocated to the human and computer in such a way that - in contrast to earlier attempts at determining operation sequences and cutting parameters automatically - it is up to the worker to plan the program's flow and to decide on the technological data. Thus, he maintains the initiative, evaluates working situations and decides on the actions to be taken. The programming system is designed to assist him in such a way that it keeps recurring geometric elements and cutting cycles available in a generalized, pre-programmed form, by which he can establish the program through setting parameters and combining the modules in the right order. Further assistance is provided by a tool data base and, most important, by graphic-real-time simulation of the tool movements as a device for testing the programs.

According to the concrete and graphic mode of thinking and acting shown by skilled workers, the user interface design is based, in constrast to common abstract and symbolic programming languages, on graphic representations of the objects and operations needed for the program and very similar to those used in drawings. In particular, the meaning of data formats that have to be filled is explained by graphic representation (see Fig. 1).

The dialogue procedures meet common requirements of robustness and self-description that always indicate to the user, in whichever system state he is in, which actions are open to him and what effects his actions will produce. They allow the user to undo a certain dialogue step without any other negative consequences. All in all, this programming system takes advantage of the performance of modern computers, in particular of its forms of interaction such as direct manipulation or the use of window techniques. It proves to be a complex, but still transparent tool for skilled workers. Thus, existing skills and knowledge are activated and maintained rather than being replaced through objectification in software (Liese 1989).

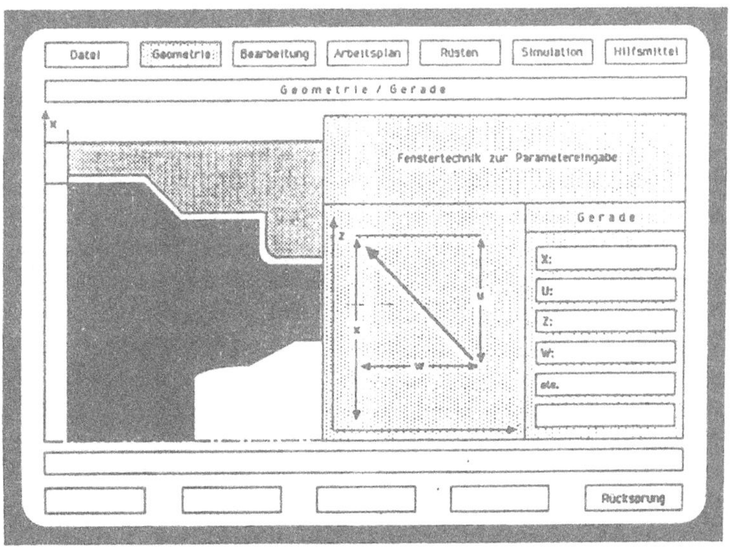

Fig. 1: Basic form of interaction (turning)

This programming system can advantageously be used at all places where NC programs are edited, complemented or corrected. Its internal functional structure and data formats have been designed in such a way that it can be integrated in and linked to existing NC and CAD technology. The system internally produces ISO standard NC code which can be further processed by any compatible controller. Considering the link to a CAD system, it is first of all important to establish adequate task allocations and responsibilities: While the designer is competent and should be responsible for the functional quality of the design, the skilled worker with his production competence should be responsible for an efficient program to produce the part. Therefore, not the designer, as in many cases, but rather the skilled worker should edit the NC-program, and for doing this efficiently the programming system provides him with the facility to call off geometric data from the CAD system he needs for the definition of the part.

4 Results and Experiences

This joint industrial R & D project with 23 partners worked very successful. The partners found very effective ways to closely cooperate in small working groups with well defined development tasks. The intensity of cooperation increased during the course of the project. Most of the partners even continued to undertake common actions after the official end of the project: they built up a joint pilot installation with the different products developed at the machine tool fair (see Fig. 2), and they operate two joint demonstration centers in Germany to promote the products and the common new philosohpy. Through their impact on the market, they even compelled the market leader to change his product strategy and to develop similar products. All this clearly indicates that the objective and scope of this development focused on a key problem for the German manufacturing industry.

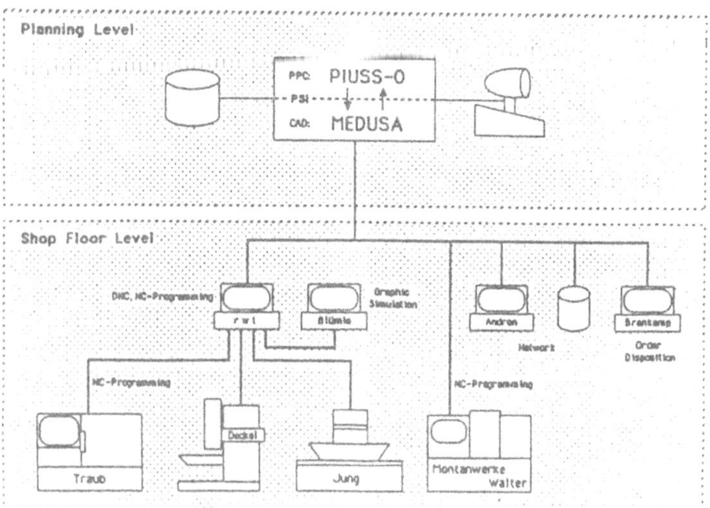

Fig. 2: Shop floor programming with data links to CAD and PPC

The prototypes being developed have been broadly tested under real production conditions with the user company, a well known producer of Diesel engines and jet turbines. More than half of its part production is organized in production islands according to the principles of group technology. The skilled workers, working in groups within those islands, are responsible for the operation and routine maintenance of a number of NC and conventional machines, for quality control, for tool preparation, and for NC programming. Against this background, the new shop floor oriented programming methods have been tested in competition to a number of existing "conventional" NC programming methods. For the case of turning, the economic superiority of the new method is shown in Fig. 3 in terms of total costs for programming and testing the programs.

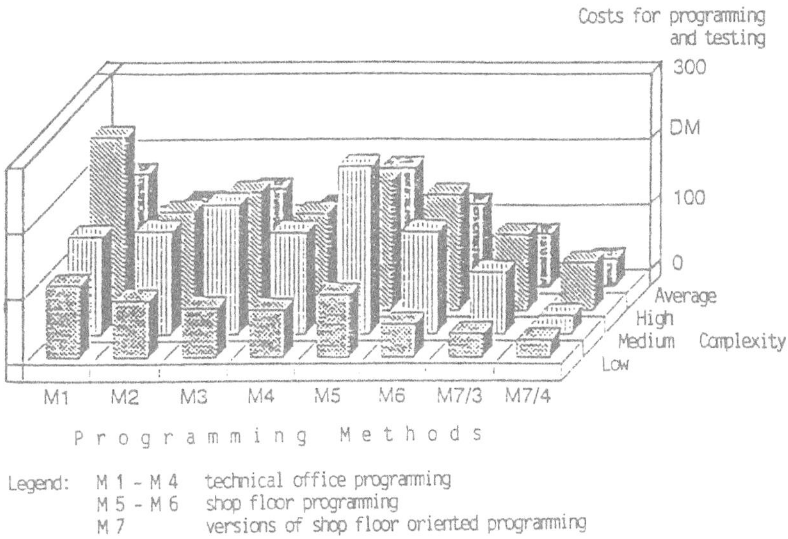

Fig. 3: Comparison of costs for programming and testing (Source:mtu)

Moreover, the suitability of the shop floor oriented programming methods as an easy-to-use and effective tool for skilled workers is demonstrated by the very favourable learning curves shown in Fig. 4. Thanks to the self-descriptive interaction procedures based on the conventional graphical "language" of the shop floor, instruction courses are no longer needed. The system can be used unaided after instruction lasting not more than a quarter of an hour. Even workers with no programming experience perform, after a five-day learning phase of learning by doing, as well as their most experienced colleagues did at the beginning.

Average time (min) for programming a part

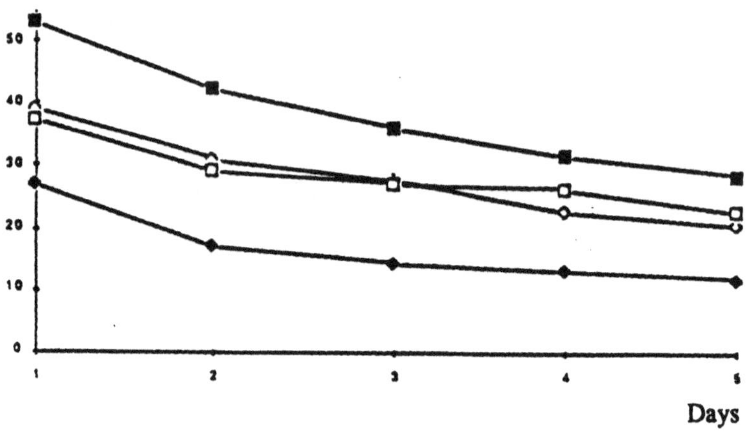

Legend:

- ♦· NC operators with programming experience
- ◇ Normal NC operators
- ■· Conventional operators (no NC experience)
- □· Trainees

Fig. 4: Learning experiment with shop floor oriented programming (Source: mtu)

5 Concluding Remarks

The history of development and use of NC technology in Germany and the economic sucess of the shop floor oriented programming methods being shortly reported here underline the relevance of industrial cultural factors for the shaping of production work and technology. The role of skilled work in German factories and its value for maintaining competitiveness under the changing conditions of global markets led to a fundamentally different shape of technology being used in a different context of work organisation. The Tayloristic tradition and the technology-centred perspective to replace human skills have been recognized as highly questionable and problematic to maintain and develop production competence and competitiveness. Technology and automation are not the solution to poor performance of an industry, they are part of the problem. And more of the same remedy only produces more of the same misery.

This case demonstrates, how important it is to choose the right perspective in system design. Technical systems cannot be designed appropriately out of the context of work design and without reflecting the mutual relationships between work organisation, skills and qualifications, and functional requirements for the technical system. Human work should not be seen in terms of failures and costs, but rather as the primary source of productivity. Instead of searching for the potentials of technology and the limits of man, efficient production might require looking for new ways of combining the unique capabilities of humans with the performance of machines. Therefore, compared with traditional views, the perspective in system design has to be shifted from developing computer programs imitating and replacing human expertise towards designing computers as a useful tool for human experts.

References

Böhle, F. and Milkau, B.: Vom Handrad zum Bildschirm. Eine Untersuchung zur sinnlichen Erfahrung im Arbeitsprozeß, Frankfurt: Campus 1988

Brödner, P.: In Search of the Computer-aided Craftsman, AI & Society (1989) 3, 39-46

Brödner, P.: The Shape of Future Technology. The Anthropocentric Alternative, London: Springer 1990

Diekmann, Th. and Klotz, U.: Veränderung der Organisation des Arbeitsablaufs bei Werkzeugmaschinen durch den Einsatz von Mikrocomputern, Forschungsbericht HA 80-050, Bonn: BMFT 1980

Kern, H. and Schumann, M.: Das Ende der Arbeitsteilung? Rationalisierung in der industriellen Produktion, München: Hanser 1984

Landesreport Baden-Württemberg Nr. 15: CIM in den USA. Evolution mit dem Ziel einer "arbeiterlosen Fabrik", Januar 1987

Lay, G., Boffo, M. and Lemmermeier, L.: Beurteilung der Wirtschaftlichkeit von CNC-Drehmaschinen unter organisatorischen Gesichtspunkten, KfK-PFT 72, Karlsruhe: Kernforschungszentrum 1983

Liese, S. (Hg.): Neue Wege der NC-Programmierung und Fertigungsorganisation, KfK-PFT 58, Karlsruhe: Kernforschungszentrum 1983

Liese, S. (Hg.): Werkstattorientierte Programmierverfahren (WOP). Ein Beitrag zur Weiterentwicklung qualifikationsorientierter Produktion, KfK-PFT 138, Karlsruhe: Kernforschungszentrum 1989

Noble, D.F.: Forces of Production. A Social History of Industrial Automation, New York: Knopf 1984

Nyhuis, F. and Nyhuis, P.: Rüstablaufanalysen an Bearbeitungszentren, HGF-Kurzberichte, Industrie-Anzeiger 108 (1986), 37-38

Rauner, F. and Ruth, K.: Perspectives of Research in "Industrial Culture", in: Karwowski, W. and Rahimi, M. (Eds.): Ergonomics of Hybrid Automated Systems II, Amsterdam: Elsevier 1990

Rempp, H. and Lay, G.: Werkstattprogrammierung bei CNC-Steuerung als Beitrag zur Aufgabenbereicherung, Rationalisierung 32 (1981), 279-281

Shaiken, H.: Neue Technologien und Organisation der Arbeit, Leviathan 1980, 190-211

Sorge, A., Hartmann, G., Warner, M. and Nicholas, L.: Mikroelektronik und Arbeit in der Industrie, Frankfurt/New York: Campus 1982

Spur, G. and Potthast, A.: Grafisches Simulationssystem für die NC-Drehbearbeitung, ZwF 76 (1981), 387-390

Stute, G. (Hg.): EXAPT. Möglichkeiten und Anwendung der automatisierten Programmierung für NC-Maschinen, München: Hanser 1969

Taylor, F.W.: The Principles of Scientific Management, 1911

v. Zeppelin, W.: Parallelprogrammieren und graphische Simulation erschließen weitere Möglichkeiten der Werkstattprogrammierung, ZwF 76 (1981), 355-359

The Social Shaping of Machine Tool Design and Manufacture in the United Kingdom: Some Preliminary Findings

by J. Martin Corbett

1 Introduction

This report outlines the major findings of a survey carried out between March and July 1991. The survey, larger scale variants of which were carried out by other CAPIRN members in a number of other machine tool producing countries within Europe (as well as in Japan, the US and the USSR), focuses on 15 UK machine tool manufacturing companies. Results from two larger UK surveys of machine tool production and use (Metalworking Production, 1987; Machine Tool Technologies Association, 1991)) are also included in this report and offer invaluable contextual information and statistics to support the interpretation of the more detailed survey.

2 Background to the Survey

In 1987 (the most recent year in which a detailed nationwide machine tool survey was carried out), the machine tool population in UK industry totalled 745,983 machine tools. 54,832 machines (7.35%) of this total were of the CNC or NC type (unfortunately, UK surveys do not differentiate between NC and CNC machine tools).

Whilst C/NC machines represent a small percentage of the total number of machines currently being used by UK industry, the actual number of new C/NC machines purchased in the UK each year represents a far greater percentage. For example, in 1990, of the £973.1 millions spent by UK industry on new machine tools, C/NC machines accounted for approximately 39% (by value) of this total investment (i.e. £383 millions). Indeed, whilst the average age of machine tools used by UK industry is 12 years, 70% of all the C/NC machines in use in 1987 were less than 5 years old.

The figure of £973.1 millions represents the apparent consumption of machine tool consumption by UK industry in 1990 (comparative figures for the US, Japan and Germany are £2662 millions, £4276 millions and £3853 millions, respectively). Imported machine tools account for £523 millions of the total UK machine tool consumption in 1990. Of the importing countries, Japan and [West] Germany have consistently accounted for approximately half of the imported machine tool population (by value). Interestingly, about half of the machine tools imported from Germany were non-CNC whilst 80% of the machine tools imported from Japan were of the CNC type (by value). In terms of numbers, of the 2907 machine tools imported from Japan, 1425 (49%) were CNC and of the 10945 machine tools imported from [West] Germany, only 946 (9%) were CNC.

In 1990 (as with every year since 1984), the UK imported more machine tools than it exported. Whilst machine tools to the value of £489.5 millions were exported in this year (30847 non-CNC machines and 4349 C/NC machines), imports totalled £523 millions (30852 non-CNC machines and 4378 C/NC machines).

As table 1 reveals, the largest export markets for UK machine tools are the USA and West Germany. Note that, over the last three or four years, EC countries have become increasingly important trading partners for both imports and exports of machine tools.

Table 1: Machine Tool Exports 1979-1990 by Major Destinations (in £ million)

	1979		1980		1981		1982	
	All Countries	222.8	All Countries	289.8	All Countries	281.2	All Countries	273.1
	EEC	52.1	EEC	75.0	EEC	58.2	EEC	54.5
1	U S A	45.0	U S A	58.0	U S A	75.5	U S A	68.5
2	Poland	15.8	W Germany	25.3	South Africa	24.8	South Africa	19.2
3	W Germany	14.9	South Africa	20.6	W Germany	21.0	W Germany	14.6
4	Italy	10.5	France	16.3	France	14.9	Mexico	13.8
5	France	10.3	Italy	12.4	Canada	13.3	France	13.6
6	South Africa	9.2	Canada	10.8	Irish Republic	7.2	India	12.4
7	U S S R	8.4	Switzerland	10.6	Italy	7.0	Irish Republic	9.7
8	Canada	7.5	Irish Republic	10.3	Switzerland	6.6	Canada	7.4
9	Irish Republic	7.0	Mexico	8.8	India	6.6	E Germany	7.3
10	South Korea	6.5	U S S R	8.3	U S S R	6.3	Italy	6.5
11	Sweden	5.5	India	6.8	Mexico	6.0	Sweden	5.5
12	Switzerland	4.8	Sweden	6.6	Nigeria	5.7	Switzerland	5.4

	1983		1984		1985		1986	
	All Countries	210.3	All Countries	226.3	All Countries	265.9	All Countries	269.2
	EEC	49.9	EEC	53.5	EEC	65.3	EEC	87.3
1	U S A	38.6	U S A	59.8	U S A	68.4	U S A	64.4
2	W Germany	17.2	W Germany	20.7	W Germany	25.4	W Germany	30.6
3	South Africa	15.7	France	11.4	Canada	14.7	France	18.8
4	France	15.7	U S S R	10.7	France	13.3	China	12.4
5	U S S R	9.6	South Africa	9.8	China	10.7	Canada	11.7
6	India	9.5	India	8.2	India	8.9	Italy	10.9
7	Mexico	7.8	Canada	7.3	Australia	7.5	India	8.9
8	E Germany	7.8	Sweden	6.3	Sweden	7.2	Australia	7.9
9	Sweden	7.3	Italy	6.2	U S S R	7.1	Sweden	7.3
10	Italy	4.6	Switzerland	5.2	Switzerland	6.2	Switzerland	7.2
11	Canada	4.6	Belgium	4.7	Netherlands	5.9	Netherlands	6.9
12	Switzerland	4.4	Egypt	4.3	South Africa	5.6	Irish Republic	6.0

	1987		1988		1989		1990	
	All Countries	305.7	All Countries	386.2	All Countries	398.6	All Countries	489.5
	EEC	119.5	EEC	146.3	EEC	171.9	EEC	236.4
1	U S A	59.4	U S A	62.2	U S A	65.1	U S A	76.7
2	W Germany	32.2	W Germany	40.4	W Germany	46.9	W Germany	74.7
3	France	21.0	Iraq	31.5	Belgium	35.3	Belgium	48.0
4	Spain	17.9	France	31.2	France	26.8	France	39.9
5	Italy	15.8	Italy	20.7	Italy	21.1	Italy	25.1
6	Canada	14.5	Belgium	18.4	Spain	14.6	U S S R	18.3
7	Belgium	12.9	Spain	12.6	Irish Republic	13.3	Sweden	16.0
8	Sweden	11.2	China	10.1	Australia	12.4	Irish Republic	12.6
9	Australia	9.1	South Africa	9.7	Iraq	12.4	Spain	12.1
10	U S S R	9.0	Australia	9.3	Sweden	10.1	Netherlands	11.0
11	Netherlands	8.2	Irish Republic	8.6	China	9.5	Switzerland	10.5
12	Irish Republic	7.2	Canada	8.3	South Africa	9.1	South Africa	9.3

Source: HM Customs & Excise

Note: Figures include export of new and used machines

Fig. 1: UK CNC Machine Tool Sales (1990).

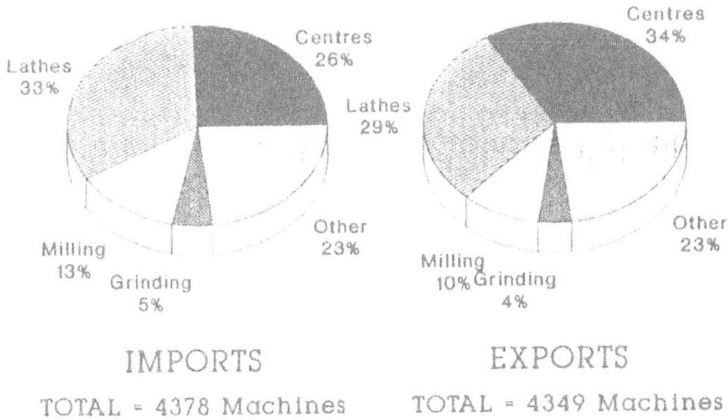

Figure 1 details UK machine imports and exports in 1990 by type of CNC machine. Note that Japanese imports account for 38% of the 4378 CNC machines imported to the UK in 1990, whilst Belgium was the biggest single market for CNC exports (17% of total) in this year.
Figure 2 provides similar data for non-CNC machines.

Fig. 2: UK non-CNC Machine Tool Sales (1990).

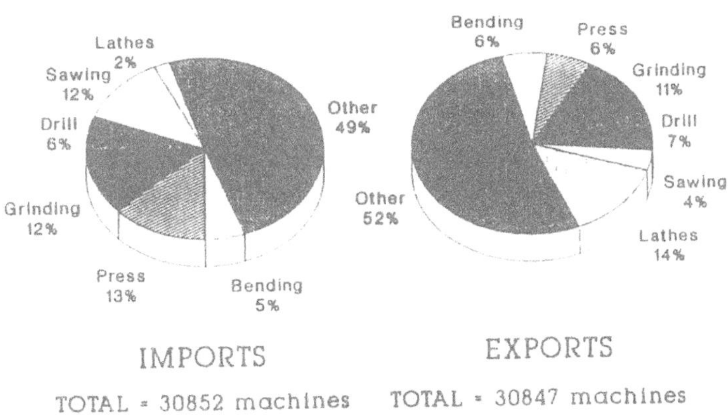

When we consider the sales of UK machine tools in 1990, it is clear that CNC machines are the growth market - with CNC machining centres and lathes accounting for the bulk of sales in terms of value, but non-CNC machines predominating in terms of actual numbers of machines manufactured and sold (see table 2). Note that the figures in table 2 are based on a national survey of approximately 60% of the UK machine tools industry. According to this survey, total sales of CNC machines totalled £360.1 millions and sales for non-CNC machines were £201.3 millions. Parts and accessories accounted for the remaining £133.6 millions.

Table 2: Sales by UK Machine Tool Manufactures. (1990). (Total = £ 695 m.).

Machine Type	No.	Value
C/NC Centre	1778	£117.2 m
C/NC Lathe	1493	£77.3 m
C/NC Milling	722	£47.8 m
Grinding	4419	£14.5 m
Drilling	1027	£2.7 m
Lathe	5744	£52.2 m
Press	534	£24.7 m

3 The Capirn (UK) Survey

The UK survey was distributed to all major UK machine tools manufacturers. Whilst there are some 65 machine tool manufacturers in the country, only 17 manufacture and sell British designed and built products and it was these 17 firms that formed the basis for the survey.

15 companies completed and returned the questionnaire (although two companies supplied a catalogue of their products rather than fill in the relevant sections of the survey). Of these 15, nine specialised in the manufacture and distribution of CNC machines. The average number of people employed by companies in the survey was 208. All companies in the survey had been in the machine tool business for between 5 and 30 years (average = 19 years).

3.1 Employment by Function

In terms of numbers employed within each function of the surveyed companies, an average of 62% of employees were directly engaged in production, 10% in production planning, 11% in marketing and sales and only 3% in research and development.

Of this 62% of employees directly engaged in production, 53% were classified as skilled craftspeople. Within machine tool-using companies in the UK, skilled craftspeople represent only 27% of personnel engaged directly in production (Machine Tool Technologies Association, 1991)

3.2 Products

In 1990, machining centres, lathes and milling machines made up the bulk of UK manufacturers CNC machine tool sales. Within the survey sample of nine C-NC-producing companies, 67% manufactured CNC lathes, 33% manufactured machining centres (vertical and horizontal) and 33% manufactured milling machines. Exports accounted for over half of the survey companies' sales, reflecting a fairly consistent trend in the UK machine tools industry regarding new orders (see figure 3). Note that figure 3 indicates the difficulties that have been experienced by the UK industry over the last twelve months in the home market. The decline in export orders is also significant - not least because it was only the growth in this area that prevented a fall in UK machine output in 1990. Interestingly, national statistics reveal that 87.5% of all UK machine tool exports (by number) are of non-CNC machines. Lathes accounted for 14%, grinding machines for 11% and drilling machines for 6% of this total.

Fig. 3: Quarterly Volume Index of New Orders for UK Machine Tool Manufacturers.

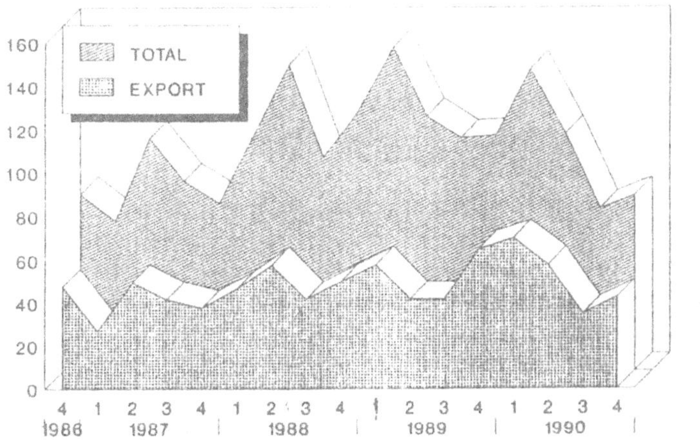

(Total new orders : 1985 = 100)

4 CNC Controllers

Only two of the surveyed companies manufactured their own CNC controller (for grinding machines in both cases). All CNC manufacturers (N = 9) offered their CNC machines with imported controllers fitted. 7 (78%) of the companies offered the FANUC 0 or the FANUC 11 series of controllers either as standard or as an option; 3 (34%) of companies offered the General Electric 2000 series controller as standard; and 2 (22%) offered the Heidenhaim TNC 151 or 155 controller.

Of the two companies offering their own CNC controllers, one offered a fairly basic interactive menu-driven interface with no graphics capability (plus a conventional G code programming facility), the other offered a sophisticated graphical interface for its range of specialised CNC grinding machines. Both were intended for use by either a skilled worker or a specialist office programmer.

Survey respondents perceived a market demand for 'user-friendly' menu-driven graphical interfaces and felt that FANUC, Yamasaki, GE and Philips had all but cornered this market. No company had plans to develop their own 'off-the-shelf' CNC controller in the forseeable future.

All controllers (domestic and imported) offered at least basic provision for DNC. The companies argued that DNC sophistication was not an important priority because of its lack of use in most UK factories. Indeed, a recent survey (Industrial Computing, 1990) shows that less than 8% of companies utilising CNC production technology use DNC linkage in production. 26% of UK users use a physically separate office-based CAD suite to produce NC tapes or CNC programmes. The remainder rely on manual data input on the shopfloor (although not necessarily by shopfloor personnel).

4.1 Machine Tool Design

In response to the question 'how would you assess the relative significance of the following statements with respect to successful/optimum machine design', a clear (and statistically significant) rank ordering was apparent. Namely,

1. Fulfil technical requirements;
2. Meet market requirements;
3. Fulfil user requirements;
4. Reduce direct labour costs;
5. Take over operational decisions;
6. Increase user capabilities and skills;
7. Take over physical work.

This rank order, with its emphasis on technical and market requirements, is similar to those obtained from the results of the Japanese and German machine tool producer surveys. Note however, that unlike German producers, the UK and Japanese sample give a low priority to increasing user capabilities and skills.

Among UK machine tool manufacturers, the main emphasis would appear to be on cost-effective and reliable hardware rather than technical sophistication and software. Non-CNC technology is often seen as the market niche most appropriate for UK machine manufacturers. All non-CNC manufacturing companies in the survey stressed the importance of overseas (as opposed to domestic) sales markets.

Table 3: UK CNC Installations by Company Size (1987).

Company Size	No. of Machines	% of UK Machines
1000 +	7238	13.2%
500 - 999	4715	8.6%
100 - 499	12940	23.6%
50 - 99	6415	11.7%
Under 50	23524	42.9%
TOTAL	54832	100.0%

CNC machine tool manufacturers appear more concerned with the domestic market and a clear pattern emerged from the survey (and a number of follow-up telephone interviews) that they perceived small companies (under 50 employees) as a key market niche for their products. Certainly, the 1987 national survey (Metalworking Production, 1987) revealed that companies of this size use 43% of all the UK CNC machine tool population (see table 3). A large majority of UK sub-contracting companies employ less than 50 people and these companies use approximately 22% of the total UK CNC machine tool population. Given this emphasis on the domestic market, and given the relatively cheap cost of direct labour in UK manufacturing, the de-emphasis on reducing labour costs (traditionally a key criterion in UK technical investment) is an interesting but clearly rational strategy.

The most dramatic market growth has taken place in the acquisition of CNC milling, turning and machining centre machines by sub-contractor companies employing less than 20 people. This size company accounted for just over 50% of all CNC machines installed by sub-contractors in 1987.

4.2 Participants in the Machine Tool Design Process

In response to the question 'how would you assess the importance of the following participants/institutions in the development of your company's CNC system(s)', the same rank ordering was apparent for both the UK companies who had developed and marketed their own CNC control systems. This rank ordering is outlined in table four below and includes results from the Japanese and German surveys for comparision. Note that the differences between these three rank orderings are not statistically significant.

However, table four reveals a distinct difference between UK manufacturers and their Japanese and German competitors in relation to the importance of trade fairs and exhibitions in CNC systems development. UK survey respondents argued that such fairs were of far more importance to sales and marketing than to research and development.

Again, for UK producers, cost-effectiveness in terms of competing with German and Japanese CNC systems was seen as of paramount importance. Yet the companies stressed the importance of a market niche based on customised CNC system configurations to suit the customer at a competitive price. Neither company had plans to invest significant sums of money for the development of more sophisticated CNC controllers. Competition from FANUC, GE and Heidenhaim (amongst others) was seen as simply too fierce to warrant such a strategy.

Table 4: Rank ordering of importance of participants in CNC machine tool design - an international comparison

DESIGN PARTICIPANTS	UK	Japan	Germany
Competitors;	1	2	2
Informal customer-producer relations;	2	4	3
Customers;	3	1	1
Suppliers;	4	7	6
Consultants;	5	8	7
System specialists;	6	5	8
Research institutes and universities;	7	6	5
Exhibitions/trade fairs	8	3	4

Those companies who fitted imported CNC controllers to their own machine tool hardware either did not answer this question or perceived only suppliers and customers as having any influence of CNC controller design.

4.3 Design Philosophy: Projected Users

The nine CNC machine tool producers questioned in the survey tended to view a skilled worker as the main user of their machines. Indeed, the image of the potential users held by UK producers was similar to that held by German CNC producers (see table 5). This reflects the fact that a majority of domestic CNC-purchasing companies employ under 50 people (see table 3). Such companies appear less prone to the bureaucratic division of labour between programmers and shopfloor machinists in evidence in larger machine tool using companies.

Table 5: Projected users of CNC Machine Tools.

USER	UK	Japan	Germany
Skilled worker	55%	13%	54%
Specialist programmer	22%	27%	15%
Semi-skilled	11%	2%	11%
No special group	22%	18%	15%
Work groups	0%	11%	3%
Residual	0%	29%	2%

The two CNC controller producers had no particular user group in mind. Note also that whilst all other companies offered identical Japanese (Fanuc) CNC controllers for their machines, they differed in their perception of who would use this controller. This clearly begs the question as to the extent to which CNC machine tool design philosophies determine user practice.

One apparent anomaly is thrown up by these data. Whilst the survey findings revealed that CNC machine tool producers give a low priority to increasing user skills and capabilities in the achievement of optimum machine design (see discussion in Machine Tool Design section above), they nonetheless tend to view skilled workers as the ideal user of their machines. None of the surveyed companies viewed their machines as tools for skills development. Rather they saw them as tools for increasing the productivity of skilled users. Skills development is seen as the responsibility of the user company - although, as we shall see in the next section, 27% of the machine tool producers do offer some form of advanced user training (for an additional cost).

Training Offered by Machine Tool Manufacturers

13 out of the 15 (87%) surveyed companies offered basic training in addition to the traditional after-sales services and guarantees. Of these companies, 4 (27%) offered training located on their own premises, 7 (46%) offered training on the user site, and 27% offered both services. In addition, 27% of the surveyed companies offered advanced training.

All companies had expanded their operations in recent years to offer a consultancy service. All but one company (93%) offered this service in the area of production planning and control (i.e. 'how to get the best out of your machine tools'). The two UK CNC system manufacturers offered additional consultancy services - 'Design to Manufacture' and 'CADCAM / FMS installation'. The other UK manufacturers of CNC machines (with imported controllers) all offered similar consultancy services, but these were via outside consultancy agencies.

5 Summary

There seems little doubt that the UK machine tool building industry is in decline. There is also evidence that UK industry invested in fewer machine tools in 1990 than in the previous year. Indeed, the UK was one of very few countries to suffer a fall in demand for machine tools in 1990. Economic recession hit in the UK before many other countries, although it has since spread and the figures for 1991 will probably reflect what has now generally become a global recession. High interests and strong overseas competition - particularly in the CNC market sector - have taken their toll over the past seven years.

In addition, almost all the major foreign machine tool builders have a factory and/or a distribution centre within the UK to take advantage of its relatively cheap labour force. This also makes any detailed examination of any UK-specific cultural analysis of the shaping of machine tool design difficult. Even UK-government funded surveys have now given up attempting to define what actually constitutes a British machine tool owing to the large number of machine tools that are designed in (say) Germany and Japan but manufactured and distributed within the UK.

The survey results, and particularly those from the large national survey of machine tool usage, also call into question the diffusion of 'flexible specialisation' or 'post-Fordism' within UK manufacturing. Small sub-contracting firms and manufacturing companies employing less than 100 people utilise over half of all the CNC machine tools currently operative within the UK. Large manufacturers would appear to be staying with dedicated machine tools and transfer devices, and use a mere 13%

of the total UK CNC machine tool population.

The flexibility of CNC has still to be exploited on the shopfloor of these larger manufacturing organisations - higher management preferring to sub-contract out the more time-consuming and skilled work (such as die-sinking) to smaller specialist companies. Whilst this trend may change as large organisations look to replace their ageing machinery and as CAD/CAM develops, a move towards large-scale 'flexible specialisation' is not discernible at the present time.

References

Industrial Computing (1990) 'The British Industrial Computing Survey'. Industrial Computing, October 1990, 31-50.

Machine Tool Technologies Association (1991) Machine Tool Statistics 1991. London: MTTA Press

Metalworking Production (1987) The Sixth Survey of Machine Tools and Production Equipment in Britain. London: Morgan-Grampian.

Conclusion: Understanding Competitiveness

by Lauge Rasmussen and Felix Rauner

Responsibility for the scientific explanation of competitiveness - 'competitive: advantage' is traditionally ascribed to economics. The success of products on increasingly global markets is primarily determined by prices, distribution systems, marketing and advertising. Secondly, it is the capacity of an enterprise to realize the transition from a seller's market to a buyer's market: delivery (supply) reliability, product quality and maximal consideration of customer wishes (flexible specialisation).

It is presumed here, that the products are globally disposable, and that they are simply made available to the customer according to the latest - "state-of-the-art" technology, with the variety required by the customer, preferably immediately and with a minimum of after-sales-service - all customer orientated. Differences in the technical realization of the products -i.e. in the shape of the technology- are interpreted as an expression of lack of simultaneity from an industrial-sociological perspective. According to this, competitive advantage is held by those who are ahead in temporal terms as regards modernisation in the development of products.

A linear process of higher development is assumed in technological development. This is the first fundamental misconception.Some economic theorists,however,have recognized the term **economic culture**.For example Peter Berger (1987)[1] and Michael Porter (1990)[2] focus on **context variables** such as 'factor conditions','demand conditions','related and supporting industries' and 'national environment'.This is the strength of their approach.The weakness is their focus on the '**rate**' and '**speed**'of development rather than on the **content** of factors like technology or skills.Therefore they are not able to explain why engineers design and customers buy a certain type of machine and not the other possible alternatives.
Our concept differs at precisely this point from Berger's and Porter's **economic culture**.The articles in this book show the necessity of understanding the **structure** as well as the
content of the competitive forces we agree with Berger and Porter that the context variables are relevant in contrast to many **organizational culture** approaches, which are mostly limited to the internal attributes of a single firm.But our dialectical approach differs from Berger's and Porter's **functionalistic** approach in the way the **contrasts** between different **institutional** forces and actors (groups,individuals) behind innovations are focused on.

Common to all approaches in this book is the search for mediating, action-directed links between the macro-orientated concept of economic culture and the micro-orientated one of organizational culture. For example, Klaus Ruth developed a model of a 'socio-technical action system'. Martin Corbett, Jürgen Friedrich, Thomas Binder/Klaus T. Nielsen and Eckehard F. Moritz on the other hand focus on 'designer/user' and 'producer/user' interrelationships as important differences between Scandinavian, German, American and Japanese design cultures. Lauge Rasmussen further discusses the relationship between institutions and actors from a dialectical viewpoint. This perspective, he explains, helps us to recognise that every action has a tendency to produce a movement in the opposite direction. According to Rasmussen, industrialization is like a multi-dimensional continuum, where the scope as well as the content of role expectations and regulations constitute conflict, as well as relatively homogenous situations. But this process, he points out, is dependent on actors for which institutions provide a background of role expectations and patterns of behaviour; There is also a foreground, where actors are capable of shaping and innovating. According to our dialectical viewpoint, every institution must change to some degree in order to survive as a fundamental cultural pattern (accommodation). If not, it may collapse and thus change in a more radical sense. The changes are initiated either by **internal** contradictions or tensions in a given culture or by **external** pressure from other industrial cultures. The tensions generate experimental action with alternatives, which is the starting point for transcending the existing state of rules or patterns of behaviour.

In certain periods the tensions are so strong that experiments become highly divergent. New and old practices are mixed together or compete with one another. Various options and alternatives are considered. Some of the existing rules and competences, including technology and skills, may be questioned. But precisely because this occurs, the more fundamental part of the institution -such as values and objectives- may be able to survive.

In other periods, convergent forces may become dominant and limit the space for alternatives until one or a few 'solutions' survive. Nevertheless, in the convergent periods the seeds of future divergent experimentation lie latent shaping the present state of seemingly more and more uniform patterns of behaviour, skills and technology. To a certain degree actors are tied to the institutions either by 'legitimate order' (Weber) or by force. The more tensions or contradictions between or within the existing institutions, the more likely it is that some actors will try to transcend the existing patterns and shape new patterns of organization and new kinds of technology skills.

The question why, for example, the American machine tool sector dropped from its leading position of "world champion" at the end of the 1970's into relative insignificance, cannot be adequately explained by economics. Obviously, the conglomeration policy propagated by the Harvard Business School played a role in this, as Gabriele Laske has pointed out in her approach in this book.
More comprehensive explanations are however necessary to explain the extent of the disaster:
In the self critical and startling MIT-Study "Made in America",[3] the manner and fashion of how skilled workers are trained in the USA is criticized as a decisive handicap:

"The American system of on-the-job-training is called following Joe around, and it does not work"[4]. This is a clear reference to the significance inherent of characteristic differences between diverse industrial cultures in the explanation of competitive advantage. Scientific approaches that over the past few years have focused on the cultural phenomenon anew - economic culture, competitive advantage of nations and enterprise culture, are succeeding in deepening the comprehension of the concept of competitiveness. As every form of competition ultimately takes place on the concrete level, dealing with concrete products, services and protagonists, an answer must finally be found to the question of why a customer, who for instance bought machine tools from Cincinnati Millicron until 1978, prefers machine tools from other manufacturers from 1980 onwards. And if this decision is based to a significant degree on the difference between the products, then a second question becomes equally significant:What influences the designer to shape products in a different way? The concept of industrial culture enables the incorporation of these questions into deliberations about competitive advantage in order to find answers to them.Seeking explanations of the above mentioned questions our international and intercultural study of machine tool manufacturing has proven to be surprisingly effective. We shall summarise our findings and insights into four theses under the perspective of "understanding competitiveness".

As an introduction we also want to answer the often posed question of 'why didwe choose the machine tool sector as an example for our research, when the economic weight of this sector is rather modest? Apart from many coincidences inherent in the establishment of international network projects, this was above all due to the judgement in 'Made in America' summarized fittingly into two simple sentences:

"It is not possible to have world class manufacturing without world class tools. It is not possible to have world class tools if one has to import them all".[5]

An industrial nation that can no longer produce its own tools, put its manufacturing trade under threat. Hence, in the case of machine tool production, we are concerned with one of the most sensitive key industries in international competition.

1. Understanding competitiveness in the context of industrial culture means that the products involved in the competition must always also be understood as the expression of different industrial design cultures.

For example, Jürgen Friedrich refers to a cross- cultural study of the orientations of software system design in Canada and Denmark. The study indicates that Canadian software designers showed a significantly stronger orientation towards technical values and a weaker orientation with respect to social-political values than the Danish designers. Eckehard F. Moritz has discerned different types of problem-orientation between the American and German designers.

"Whereas the American designers devote most of their attention to reaching the goal, German engineers will more likely concern themselves with the understanding, formulation, and definition of the problem itself. They want to understand the problem comprehensively and from there on, to concentrate on the potentially best approach - a formidable means of developing technologically superior designs, but with the fatal probability of overshooting the mark or entirely losing sight of the intended goal."[6]

This finding is also supported by G. Laske's investigation:

"Germans have the tendency to develop machine tools not so much as optimum and low-priced production tools, but as some kind of technical works of art. This approach towards technology development is primarily interested in a product's technological aspects and much less in the practical and environmental side of its application. It is expensive and time-consuming, because users' criticisms and corrections are then received at a late development phase of the first prototype. Changes at this stage increase prices of machines more than at an earlier stage.
Therefore it is not surprising that German CNC machine tools are characterized in the United States as being overengineered. Therefore, German CNC machine tools are only sold to niche markets, but standardized German machine tools have no chance of succeeding in this area."[7]

In Japan NC-technology has been developed, from the beginning, in terms of reliable, low cost, easy-to-use machines that serve the needs of big and small enterprises alike.

As Peter Brödner points out:

"Available figures show that from the late seventies onwards, almost two thirds of inner-Japanese NC-machine sales have gone to smaller enterprises with less than 300 employees. The use of NC technology was combined with skilled work, where technological skills and technology could develop together. This fastly growing demand induced, in turn, the production of standard NC machines. With exports not higher than 30 %, the rise of the Japanese machine tool industry can only be explained by these internal Japanese market relations and development conditions."[8]

More examples of the influence of different design cultures on the type of technology developed can be shown, but the above mentioned are sufficient to represent the relevance of our general thesis (1.). We believe that the aspect should be investigated further as suggested in the next section of this chapter.

2. During the divergent phase of development, competitive advantages decisive in the market sphere are achieved by the competitors of comparable products because of certain features of the industrial design culture of nations and regions. The international quality competition during this phase filters out those technological development lines, that:

- *are best adapted to diverse industrial cultures or*

- *that are adaptable to the different industrial cultures.*

For example, David Noble showed how different approaches to the later dominating Tayloristic NC-approach were developed in the US in 1947/48. Similarly, Rauner and Ruth point out that:
"In the beginning period of CNC history many different concepts were developed... but the middle of the 1980's was the turning point after which converging forces became more influential".

The strengthened position of Japanese and German machine tool producers offered new shaping possibilities that were influenced by different employment conditions in small and medium enterprises with small batch sizes, highly skilled work forces and other factors of industrial culture. On the other hand, as Gabriele Laske points out, US machine tool builders tried to develop and produce their own controls. But the electronics and controls on the American machines were not as reliable as Japanese control systems and the lack of a satisfactory US control solution drove customers to fill their needs on the international market. The Japanese lead in controls enables them to build lower-priced universal machine tools even in the area of specialized and high-precision machines.

According to Gabriele Laske: "In the relationship between German control manufacturers and machine tool builders the bias towards special, sophisticated solutions continues to be favoured. This results from R & D approaches which do not perceive technological artefacts as embedded in a social environment with specific production philosophies, workforce qualification and work organization. Rather, machine tools are seen and designed more as a brilliant work of art where electronic and mechanical engineers demonstrate their superior knowledge. This is doubtlessly a solid resource from which to develop high-tech solutions, but this is at the price of losing extensive market shares."[10]

The dynamics between divergent and convergent forces behind the development of NC/ CNC technology is demonstrated by the **cyclical** - rather than the linear - development trends as mentioned by Rauner and Ruth. This does not imply that all types of technology follow the same cyclical development path. Other kinds of technology may follow other paths depending on the particular content and structure in question. The dynamic of divergent and convergent forces is a more general trend. We believe though, that further research is needed to confirm this statement.

3. The introduction of new production technology, even where it is fitted in well into an existing industrial culture, unleashes processes of change within this industrial culture. This dialectical relationship of assimilation and accommodation refers to a dynamic equilibrium in the process of technological innovation.

For example, G. Laske mentions that the overwhelming Japanese presence in the area of basic technologies on the U.S. market puts strong pressure on existing values and norms. In the U.S. it is widely believed that the U.S. corporations have to seek new forms of organization and ways of training their workforce. As pointed out in the MIT study "Made in America":

"It will not do to borrow pieces of the (West) German or the Japanese systems and try to make them fit into an American context. Rather, for any of the reforms to succeed and flourish, the environment in which they are implanted must be transformed." [11].

Thus, the process of influence between technology and other factors of industrial culture such as attitudes, social institutions and education goes in both directions. The different approaches,that try to transfer pieces of an industrial culture from one society to another, have often proved to be unsatisfactory, in particular if the transferred piece is closely related to the specific context variables of the society in question.The dialectical relationship between assimilation and accommodation is an important analytical perspective for understanding why some kinds of transfer succeed and others do not.

During the assimilation process, a technological innovation is exposed to an existing structure and tries to adapt to it. If the existing structures represent an obstacle to the innovation then an imbalance develops in the relationship: From this, a process of structural adaptation or accommodation develops. A new equilibrium is established. Assimilation and accommodation processes in this way become the impetus for innovation.The driving force presumed behind this is the dynamic balance, to which the reciprocal process respectively tries to adapt.Assimilation as it is used here refers to a technological process of innovation in the course of which the innovation adjusts to fit into the existing structures of industrial culture or adapts to them. Typical for this is the example of NC-technology as it developed to its industrial peak in the United States during the 1960's. The NC-technology could easily be absorbed by the existing industrial structures within the spheres of the automobile and aircraft industries in the USA at the time, since NC-technology had,in a Tayloristic sense, the strict division between planning and executing tasks as a prerequisite.

Every assimilation process however is always also coupled with an accommodation process, and an accommodation process designates the process of adjustment of the existing structures to the technological innovation. This i demonstrated in our example, where not only the engineers, but also the machine operators had to learn how to use the NC-technology.

4. *Industrial developments that take place in different industrial cultures and countries, are economically and politically stable only when neither the forces of assimilation nor accommodation dominate one-sidedly.*

A country that implements a technology from another totally different industrial culture without changing it al all overtaxes the accommodation forces of its own culture as a rule. This often results in unsuccessful approaches. A very similar effect eventuates where the concept of production culture is overstretched and a technology is adapted to a specific industrial culture to such an extent that its potential stimulation is watered down. This may occur to such a degree that little or no innovation takes place.

'Understanding competitiveness' thus means that the specific dynamic industrial culture equilibrium between assimilation and accommodation must be striven for in the process of the design and introduction of technological innovations.

This fundamental insight is equally significant for global (intercultural) and national (cultural) developments in production. G. Laske's study shows in a dramatic example how the model that engineers adhered to in the United States, in the development of NC-controlled machine tools, was influenced by the fantasies of American space technology, producing machine tools that have little in common with everyday reality in the American manufacturing industry.

In the American culture until now, technological evolution means to conquer the unknown rather than to improve already applied technologies.Similar trends are observed by Martin Corbett in the UK machine tool industry. In the UK too highly sophisticated technology is also developed but the English machine tool industry is in decline, as Corbett observes. Why is this so? Because the UK and US both contain fragmented industrial cultures in which the basic means of production are given a low priority.

The communicative industrial culture is characterized by a human network core.

Yoshimi Ito stresses this point when he writes:
"... in fact, the Japanese research and development activities are activated by the well-configured human network. In other words, the high qualification of the Japanese machine tool industry rests to a large extent on the strong human network, backed up by the academic staffs of several leading universities and related organizations". [12]

This point is also stressed by Eckehart F. Moritz:

"The Japanese are absolutely superior in every kind of activity that precedes the actual design process, especially in the accumulation of potentially relevant information."[13]

German machine tool builders like to present their CNC machine tools as some kind of masterpiece, as G. Laske points out:[14] "The German designer is much less orientated towards the potential customer's actual needs, than to develop a highly sophisticated system. "She claims: This does not mean that German machine tool designers do not listen to machine tool users. Traditionally the relationship between German machine tool users and producers has been very close." However the overemphasis on this type of relationship has perhaps become a disadvantage as a German engineer explained: "German design engineers are used to listening to their customers. Actually this seems to be something positive. But - they listen to everybody. An approach such as this cannot lead to the development of simple, standardized machines."

The communicative industrial culture is characterized by a human network covering the relationship between service institutions, producers and users as well as the relationship between the different departments in the companies, in particular the

design and production departments. This does not imply that total agreement and absence of contrasts prevails in the communicative culture. Because opposing forces, too, may act in a constructive manner, as the dynamic balance between assimilation and accommodation transforms tensions into creative design and innovations related to the actual or potential needs of the users.
We believe that the Japanese industrial culture shows characteristics which are closest to the communicative aspects discussed above. But this is not the same as saying that the Japanese machine tool sector has solved all its problems. A highly communicative culture such as the Japanese design culture shows up other weaknesses, as Eckehard F. Moritz points out:

"...creating new designs, especially surprising mechanical solutions to technical problems, can still not really be considered as one of the strong points of the Japanese designers. Just as copying your master is highly esteemed in Japanese society, inventing your own way is not."[15]

We can observe once more the seeds of future dilemmas or tensions growing not only in highly fragmented cultures such as the German and American ones, but also in highly communicative ones like the Japanese.

It is a long way from 'understanding competitiveness' to the successful organization of innovation.
The social organization of innovation aims at the network of technology, work- and labour market structures and contents, just as much as at qualification and the multifariousness of involved institutions and protagonists, in short, the innovative milieu of the regions.

The exceedingly apt term "milieu", which is deliberately used because of its lack of precission, reveals how far removed the sciences are from an encompassing understanding of innovation.

If competitiveness, furthermore, is measured not only according to economic criteria, but also by criteria of social modernisation policies that encompass a multitude of social developments, then the limitations of a practical social application inter alia, become apparent. 'Understanding competitiveness' in the context of the main research emphasis of 'industrial culture' only makes sense, when the results flow into a regional, national and international social dialogue as the basis for the shaping of social modernization processes.

GPSR Compliance
The European Union's (EU) General Product Safety Regulation (GPSR) is a set of rules that requires consumer products to be safe and our obligations to ensure this.

If you have any concerns about our products, you can contact us on

ProductSafety@springernature.com

In case Publisher is established outside the EU, the EU authorized representative is:

Springer Nature Customer Service Center GmbH
Europaplatz 3
69115 Heidelberg, Germany

www.ingramcontent.com/pod-product-compliance
Lightning Source LLC
Chambersburg PA
CBHW071436230926
51846CB00001B/247
* 9 7 8 3 5 4 0 7 6 0 2 9 0 *